D0850179

ELECTROANALYTICAL CHEMISTRY

Dropping mercury electrode and silver chloride reference electrode (Courtesy IBM Instruments).

ELECTROANALYTICAL CHEMISTRY

Basil H. Vassos

Professor
Department of Chemistry
University of Puerto Rico

Galen W. Ewing

Professor Emeritus
Seton Hall University
Adjunct Professor
New Mexico Highlands University

A Wiley-Interscience Publication

JOHN WILEY & SONS

New York   Chichester   Brisbane   Toronto   Singapore

Library of Congress Cataloging in Publication Data

Vassos, Basil H.
  Electroanalytical chemistry.

  "A Wiley-Interscience publication."
  Includes bibliographical references and index.
  1. Electrochemical analysis. I. Ewing, Galen Wood,
1914–   . II. Title.

QD115.V28   1983   543'.0871   82-17400
ISBN 0-471-09028-X

Printed in the United States of America

10 9 8 7 6 5 4 3 2 1

PREFACE

In this book we have attempted to include those aspects of electrochemistry, both theoretical and practical, that we feel a graduate student specializing in analytical chemistry should master. The early chapters contain an abbreviated review of material that should be familiar from prior courses, leading directly into more specialized advanced topics. Subsequent chapters then treat the several subdisciplines in detail.

Principles of quantitation *per se* are treated in some detail in Chapter 15, with emphasis on titrimetric and standard addition methods. The intent is to tie together many of the techniques from earlier chapters, showing their relations to each other in a way that will reinforce the individual treatments.

Similarly, Chapter 16 serves to unify the theoretical aspects of the various dynamic electroanalytical methods with a more extensive treatment of diffusion phenomena. The fractional calculus is given a brief elementary exposition.

The final chapter is a short description of the electronic instrumentation used in modern electroanalytical apparatus. This material can, of course, be omitted if the student has sufficient background in this area.

We have included literature references, partly to indicate the sources for statements or quotations, and more importantly, to suggest to the inquiring student where to find further material on a particular subject.

The book is suitable for upper-class undergraduate and first-year graduate students in chemistry and allied fields such as biology and physics. A working knowledge of calculus is assumed.

We wish to thank Professor Peter E. Sturrock for a careful and critical reading of the manuscript and for many useful suggestions. We are grateful to Mr. Stanley P. Dodd of Sargent-Welch Scientific Company for the loan of a Sargent-Welch Polarograph with which a number of the illustrations were prepared. Professor Ronald D. Clark of New Mexico Highlands University kindly permitted our use of their facilities. Our thanks also go to Gayle Foss Ewing, who typed the manuscript with the aid of a word-processing computer, and Myrtle King who assisted in reading proof.

BASIL H. VASSOS
GALEN W. EWING

Rio Piedras, Puerto Rico
Las Vegas, New Mexico
August 1983

CONTENTS

Chapter 1.   INTRODUCTION   1

What Is Electrochemistry?   1
Electroanalytical Measurements   4
    Potentiometry   4
    Galvanostatic Measurements   6
    Potentiostatic Measurements   7
Electrochemical Conventions   8
References   11

Chapter 2.   ELECTROCHEMICAL MEASUREMENTS   12

Voltage measurements   12
    Voltage Measurements with Finite Current   15
Impedance Measurements   16
The Electrical Double Layer   20
Electrocapillarity   23
Current Measurements   25
Diffusion Transport   33
References   35

Chapter 3.   POTENTIOMETRY   37

Electrode Potentials   37
    Liquid Junction Potentials   39
    The Indicator Electrode   41
Classification of Electrodes   42
The Glass Electrode   44
Ion-Selective Electrodes   48
    Classification of Ion-Selective Electrodes   50
Instrumentation   58
References   58

Chapter 4.   VOLTAMMETRY: I. POLAROGRAPHY   60

Introduction   60
Voltammetry   60
Polarography   62
    The Basic Experiment   63
Theoretical Considerations   70
    The Ilkovič Equation   70
    The Heyrovský–Ilkovič Equation   73
    Irreversible Systems   74
    Anodic Oxidations   76
    Polarographic Maxima   78
Instrumentation   78
    Sampling   79
    Resistance Compensation   80
    Charging-Current Compensation   80
    Cell Design   80
    The Capillary System   83
Experimental Methodology   83
    Supporting Electrolytes and Half-Wave-Potentials   83
    Resolution   84
    Precision, Accuracy, and Sensitivity   86
    Scope of Applicability   86
    Organic Applications   87
References   88

Chapter 5.   VOLTAMMETRY: II. PULSE AND SQUARE-WAVE POLAROGRAPHY   90

A Basic Experiment   90
Classification of Step Methods   92
The Fundamental Process   94
Instrumentation   100
References   101

Chapter 6.   VOLTAMMETRY: III. AC POLAROGRAPHY   102

The Basic Experiment   102
Theoretical Considerations   104
Instrumentation   108
    Second-Harmonic AC Polarography   109

Methods Related to AC Polarography 113
AC-Pulse Polarography 114
References 115

Chapter 7. VOLTAMMETRY: IV. LINEAR SWEEP 116

The Basic Experiment 116
Theoretical Considerations 119
   Irreversible Processes 126
   Coupled Chemical Reactions 127
   Staircase LSV 128
   AC and Pulse LSV 128
References 129

Chapter 8. VOLTAMMETRY: V. FINITE DIFFUSION 130

Thin-Layer Cells 130
Electrochemistry with Immobilized Reagents 135
Membrane Electrodes 136
Ultramicroelectrodes 138
References 138

Chapter 9. CONTROLLED-CURRENT METHODS 140

Chronopotentiometry 140
   The Basic Experiment 140
   Theoretical Considerations 142
   Instrumentation 146
   AC Chronopotentiometry 146
   Chronopotentiometry with Increasing Current 147
Coulostatic Analysis 149
References 150

Chapter 10. METHODS WITH CONVECTION: I. ELECTRODEPOSITION AND COULOMETRY 152

Electrodeposition 152
   The Basic Experiment 152
   Theoretical Considerations 153
Coulometry 155
   Controlled-Potential Coulometry 156

Constant-Current Coulometric Titration   157
Instrumentation   158
References   161

Chapter 11.   METHODS WITH CONVECTION:
               II. HYDRODYNAMIC VOLTAMMETRY   162

Classification of Methods   162
    Interface Renewal   162
    Flow-through Systems   163
    Continuously Stirred Systems   163
Disk Electrodes   163
    A Basic Experiment   163
    Theoretical Considerations   165
Flow-Through Electrodes   167
    Experimental Techniques   169
    Applications   171
References   174

Chapter 12.   STRIPPING ANALYSIS   176

The Basic Experiment   176
Theoretical Considerations   177
    Preconcentration (Plating)   177
    Stripping   178
Instrumentation   179
    Electrodes   180
References   183

Chapter 13.   CONDUCTOMETRY   184

The Experimental Basis   184
    With Electrodes   184
    Without Electrodes   187
    Magnetic Induction   187
Theoretical Considerations   189
Instrumentation   189
    The Four-Electrode Cell   189
    The Two-Electrode Cell   190
    Electrodeless Cells   193
References   194

Chapter 14.
OPTICAL–ELECTROCHEMICAL METHODS

OPTICAL–ELECTROCHEMICAL METHODS   195

Spectroelectrochemistry   195
Electrochemical Photochemistry   199
Photoelectrochemistry   201
Electrochemiluminescence (ECL)   202
Surface-Enhanced Raman Spectroscopy (SERS)   203
References   203

Chapter 15.   TECHNIQUES OF MEASUREMENT

TECHNIQUES OF MEASUREMENT   205

Comparison with Standards   205
Standard Addition or Subtraction   207
Titrimetry   208
    Endpoint Detection   209
    Constant-Potential Titrimetry   214
Coulometric Titration   215
References   217

Chapter 16.
SOME ASPECTS OF DIFFUSION PHENOMENA

SOME ASPECTS OF DIFFUSION PHENOMENA   218

Diffusion Transport   218
    The General Equation   222
Semi-Integral and Semidifferential Techniques   225
    Measurement of μ and ϵ 227
References   228

Chapter 17.   ELECTRONIC INSTRUMENTATION

ELECTRONIC INSTRUMENTATION   229

Electrical Quantities   229
Components   230
Operational Amplifiers   231
Analog Modules   236
Digital Modules   236
Noise   237
    Modulation   239
    AC Filters   240
Phase Relations   242
Microprocessors   244
References   244

CONTENTS

Appendix 1. SYMBOLS AND ABBREVIATIONS   245

Appendix 2. STANDARD ELECTRODE POTENTIALS 249

INDEX   251

ELECTROANALYTICAL CHEMISTRY

Chapter 1

INTRODUCTION

WHAT IS ELECTROCHEMISTRY?

In the chemical reactions known as *redox*, the fundamental step is the exchange of one or more electrons between two species, 1 and 2:

$$OX_1 + ne^- \longrightarrow RED_1 \qquad \text{(1-1a)}$$

$$\frac{RED_2 \longrightarrow OX_2 + ne^-}{OX_1 + RED_2 \longrightarrow OX_2 + RED_1} \qquad \begin{array}{c} \text{(1-1b)} \\ \text{(1-1)} \end{array}$$

where OX and RED represent the oxidized and reduced forms, respectively. Frequently, this fundamental process is complicated by other chemical changes. For example, when bromate ions and arsenic(III) atoms are involved in a redox reaction, the process is:

$$BrO_3^- + 6H^+ + 6e^- \longrightarrow Br^- + 3H_2O \qquad \text{(1-2a)}$$

$$\frac{3As(III) \longrightarrow 3As(V) + 6e^-}{BrO_3^- + 3As(III) + 6H^+ \longrightarrow Br^- + 3As(V) + 3H_2O} \qquad \begin{array}{c} \text{(1-2b)} \\ \text{(1-2)} \end{array}$$

Even with the complications of the protons and water molecules, the basic process is simply the transfer of electrons from arsenic to bromate ions.

The importance of a redox reaction, in the present context, lies in the fact that the transfer of electrons from reductant to oxidant can be made to take place at a pair of electrodes connected through external circuitry. At one electrode (the anode), the reductant transfers one or more electrons to the metal electrode [as Eqs. (1-1b) and (1-2b)], while to maintain overall electrical balance, an equal number of electrons must leave that electrode and pass through the external wiring. Simultaneously, the cathode yields up a like number of electrons to the oxidant [Eqs. (1-1a) and (1-2a)]. This constitutes a complete electrical circuit, and the extent of the redox process can be monitored or controlled by electronic operations on the external portion of the circuit. It is this ability to control the extent and direction of a reaction by electrical means that constitutes the unique importance of electrochemistry.

Analytical electrochemistry is concerned with small currents (seldom greater than a few milliamperes) at low voltages (up to perhaps 2 volts). This is precisely the magnitude easily handled by modern integrated-circuit electronics, allowing the vast body of electronic techniques of measurement and control to be utilized directly. This explains the facility with which electroanalytical chemistry has developed into a highly sophisticated group of instrumental techniques.

An electrochemical cell can be simulated by resistors and capacitors (R, C) combined in a network that will possess nearly identical electrical behavior. Such a network, designated an *equivalent circuit*, is often useful in theoretical studies, to assist in elucidating the properties of a cell. Similarly, the actual RC-network can be used as a *dummy cell* to substitute for the real one in testing the operation of an electrochemical instrument.

In order to achieve control of the reaction, it is necessary to prevent the oxidant and reductant from coming into direct contact with each other. For this purpose, a special container provided with a membrane or barrier is generally required. An example is shown in Figure 1-1, where the redox process is $Zn + Cu^{++} \longrightarrow Cu + Zn^{++}$. In this case the zinc electrode generates an excess of electrons that will flow along the wire, pass through an external instrument, and return to the copper electrode. The electrical circuit is then completed by ions moving through the solution between the two electrodes. The "porous diaphragm" is a barrier, such as fritted glass, to prevent the mixing of the two solutions, while permitting the transport of ions.

The electrode where oxidation takes place, in this case the zinc, is the *anode*, and the electrode where reduction occurs is the *cathode*. It is convenient to speak of

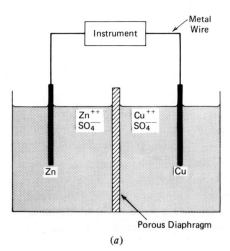

(a)

$Zn \,|\, Zn^{++} \,\|\, Cu^{++} \,|\, Cu$

(b)

Figure 1-1. (a) Electrochemical cell; (b) Symbolic representation. The anode is conventionally shown on the left.

anodic and cathodic half-reactions, while in terms of charge transport, we define corresponding anodic and cathodic currents. In practice, the direction of the reaction, and hence which electrode acts as the anode and which the cathode, depends on the nature of the external instrument. If the instrument is passive, in that it merely either allows or prevents the electron flow, the cell will show a spontaneous anode and cathode, as described above. In contrast, in many electroanalytical experiments, active instrumentation is used to control the electron flow. In this case, the electrons can be forced to move in either the spontaneous or the opposite direction; they can even be made to move rapidly in alternating directions, so that anode and cathode are periodically interchanged.†

If the cell is left without an external controlling instrument, it will exhibit a difference of potential, E, between its electrodes, indicating the tendency of the electrons to circulate outside the cell. This potential, in turn, is a measure of the free energy, ΔG, of the reaction:

$$E = -\frac{\Delta G}{nF} \tag{1-3}$$

where the negative sign indicates that, for a spontaneous reaction, ΔG is negative. The potential of an electrochemical cell is always taken as positive, and is the physically measurable difference of potential between the electrodes.‡

In Eq. (1-3), the factor n is the number of electrons exchanged, and F is the Faraday constant (estimated to be 96486.332 coulombs/mole). Equation (1-3) is strictly valid only if there is no current passing through the cell. This is because the passage of current causes not only changes of concentration, but also voltage drops and heat effects. Fortunately, with the very small currents needed by modern measuring devices, the error is usually negligible, and one can measure E, and thus ΔG, with great accuracy.

For electric current to pass through the cell, redox (faradaic) processes must occur at both electrodes, to the extent of one mole for each nF coulombs. It can thus be written:

$$\text{Number of} \atop \text{Moles reacted} = \frac{Q}{nF} = \frac{1}{nF}\int I\,dt \tag{1-4}$$

where Q is the number of coulombs passed, which, in turn, is equal to the time-integral of the current, I.

---

†Sometimes a distinction is drawn between *galvanic cells*, in which the reaction takes place spontaneously, and *electrolytic cells*, in which the direction of the reaction is reversed. This is not a particularly helpful classification, at least for us, because a given cell can be equally well operated either way.

‡On the other hand, if the potential is defined thermodynamically rather than experimentally, it is necessary to allow the reversal of the sign of E with the reversal of ΔG. Thus, if the chemical reaction is written backwards, the free energy of reaction changes sign, and it would complicate the algebra if E could not follow this change [1].

According to Eq. (1-3), in reversible systems, for every value of ΔG there is a corresponding value of the potential E of the cell. This relationship also is valid in the reverse sense, so that if one forces a potential E upon a cell, the free energy of the reaction will change by the necessary amount so that the relation $\Delta G = -nFE$ continues to hold for negligible currents. A more negative value of ΔG will cause the reaction to go forward, while a positive value will reverse the direction. The system will proceed on a path leading to the equilibrium condition.† This process is effected by changes in activities (and thus in concentrations), since ΔG depends on activities through the relation:

$$\Delta G = \Delta G^\circ + \sum_i RT \ln a_i \qquad (1\text{-}5)$$

in which ΔG° is a constant depending on the reference state, while the variables symbolized by a_i are the activities of the various species involved.

Upon changing the potential, the activities in Eq. (1-5) will begin to change until ΔG goes to zero. This occurs by means of redox processes, which alter the relative concentrations. Eventually, a form of equilibrium is attained where no further change takes place. This is not a true chemical equilibrium in the conventional sense, since upon removing the applied potential, the reactions will occur again. Chemical and electrochemical equilibria have the same properties, including reversibility, for as long as the potential is applied.

The control of ΔG by electrical means, as outlined above, is unique in chemistry. Otherwise chemists have little or no control over the free energy of a reaction once they have mixed their reagents.

ELECTROANALYTICAL MEASUREMENTS

Electrochemical processes are involved in many areas such as industrial synthesis, corrosion studies, physiological experimentation, and battery research, in addition to analytical measurements. This book is limited to those applications that can give analytical information. In other words, the emphasis is on measurements that ultimately lead to either the quantity or the concentration of chemical species.

In this context, there are three principal types of electrochemical experiments and three kinds of controlling or measuring devices to implement them. The three classes of experiments are: potentiometric, galvanostatic, and potentiostatic. We will consider them in turn.

Potentiometry

This is a type of measurement in which the function of the controlling device is primarily to ensure that no significant current is drawn from the cell. Voltage is

---

†The rate at which the equilibrium is approached varies from system to system, and for some it may be practically zero.

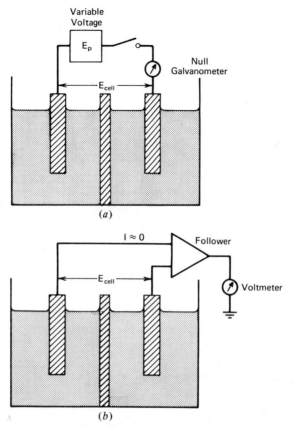

Figure 1-2. Two types of potential measurements. (a) The potentiometer; the voltage E_p is manually adjusted until exactly equal to E_{cell}. (b) The electronic voltage follower.

conveniently measured by either of two instrumental approaches: (1) the *potentiometer*, an instrument that uses an equal and opposed voltage to compensate the voltage of the cell, and (2) the *voltage follower*, which simply reproduces the voltage of the cell by electronic means. The voltage follower has become by far the dominant device; the potentiometer is used occasionally for high precision measurements. The principles of the two methods are shown in Figure 1-2.

In both cases a minute current is drawn from the cell. In the potentiometer, this is the current needed to activate the galvanometer that serves to indicate whether or not the two opposing voltages are equal. In the voltage follower, current is needed to drive the electronic circuit. In either case, a measurement usually draws less than 10^{-9} A for the brief period needed to make a reading. This corresponds to perhaps 10^{-9} coulomb, or about 10^{-14} mole of matter oxidized or reduced. This amount is so small that the concentrations are not affected, and the potential obtained is very precisely the desired equilibrium potential.

In order to obtain meaningful analytical measurements in potentiometry, one of the electrodes should be of constant potential without changing from experiment to

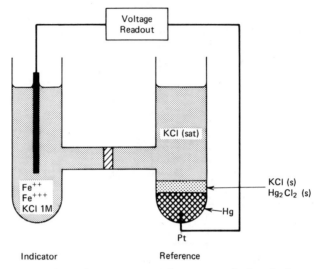

Figure 1-3. A potentiometric measurement using a saturated calomel reference electrode.

experiment. An electrode fulfilling this condition is called a *reference electrode* (Figure 1-3). Because of the invariance of the reference electrode, any change in the cell potential must be due to the contribution of the other electrode, called the *indicator* or *working electrode*.

A good reference electrode must be stable with respect to time and temperature. Its potential must not change when the current necessary to make a measurement is passed through it. Reference electrodes are available commercially in unitary construction that includes the required porous membrane.

Let us now turn to two major classes of experiment in which significant current is allowed to flow. These comprise methods based on galvanostatic operation in which the current is held constant or programmed while the voltage is measured, and potentiostatic operation, which is just the reverse.

For these methods, it is desirable to utilize a cell containing three electrodes (an anode, a cathode, and a reference electrode), so connected that the reference electrode cannot pass any appreciable current. The working electrode, where the electrochemical process that we are attempting to observe takes place, can be either the anode or the cathode. The remaining electrode is commonly designated as the *auxiliary* electrode, and we are generally not concerned with the reactions taking place in its vicinity.

Galvanostatic Measurements

If the current passing through an electrochemical cell is specified, the extent of the total redox process is completely determined at each moment. The only condition required of the chemical species is that there is enough of it present at the electrode to be oxidized or reduced at the rate required by the current. In this case

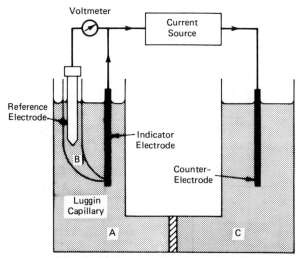

Figure 1-4. A three-electrode galvanostat. Compartments A, B, and C, may contain solutions of different compositions.

Eq. (1-4) applies, and for every nF coulombs passed there will be one mole of substance oxidized or reduced.

The experimental setup for this type of operation is shown schematically in Figure 1-4. The indicator electrode changes its potential as the electrolysis proceeds, and in order to monitor this change, a separate potentiometric measurement is made between the indicator and reference electrodes. The accurate determination of this potential is facilitated by the fact that there is very little current in the potentiometric branch (AB); essentially all the current passes between A and C.

The working electrode is the only chemical sensor in the system, since the substances in compartment C do not influence the potential measurement, and the reference electrode voltage is constant. The separating membrane can be made of fritted glass. The reference compartment is brought to a fine opening (the *Luggin capillary*) near the surface of the indicator electrode. If the tip of the Luggin capillary were further away from the surface, it would respond not only to the potential of the indicator, but also to the voltage generated by the resistance of the solution. (By Ohm's law, when a current I passes through a medium of resistance R, it generates a voltage drop $E = IR$.)

The information obtained in galvanostatic measurements is usually presented in the form of a graph of voltage against current or against time. Such a curve serves to determine concentrations, as will be shown in later chapters.

Potentiostatic Measurements

Potentiostatic experiments can be carried out using the equipment shown in Figure 1-5. The setup is quite similar to that for galvanostatic control. The basic difference is that the potentiostat automatically monitors the actual voltage and main-

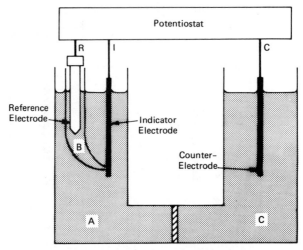

Figure 1-5.   A three-electrode potentiostat.

tains it at the desired value by adjustment of the current. The information is conveniently taken in the form of a graph of current as a function of voltage or time;
from measurements on the curve, one can calculate the analytical concentration.

Since a potentiostat can maintain the desired voltage only by manipulating the
current, there must be a limit beyond which it cannot function, simply because the
current is restricted by the capacity of the power supply. The available range is
called the *current compliance*. Typically, a potentiostat might have a current compliance of 1 A, which is more than sufficient for most types of analytical work. If
larger currents are required, the system saturates.

Similarly, galvanostatic systems are characterized by a *voltage compliance*, for
example, 10 V. The galvanostat is able to maintain the desired current only as long
as the voltage across its terminals does not exceed the compliance. If the demand
exceeds this value, the galvanostat will also go into saturation. Saturation is an
error condition and is not desirable except as a fail-safe measure.

Electrochemical Conventions

In all three types of experiments described above, the indicator electrode is the
chemical sensor, and the only chemical effects that we are really concerned with
are those occurring in the indicator compartment. Hence it is expedient to concentrate our attention on the indicator half-reaction, which can be expressed in
general terms as $OX + ne^- \longrightarrow RED$. It must always be understood that this
process cannot take place alone, but only in association with some other half-
reaction.†

---

†Free excess charges cannot exist within laboratory limitations. Guggenheim's calculations
[2] show that even if only 10^{-10} mole were to react, and if uncompensated free charges should
result, the container would be raised electrostatically to a million volts! This indicates force-
fully that the isolated half-reaction is a hypothetical construct, and not a laboratory reality.

Conventions have been established whereby a *half-cell potential* is associated with each electrode. The overall cell voltage can be calculated for a two-electrode cell by taking the difference between the two half-cell potentials:

$$E_{cell} = E_2 - E_1 \qquad (1\text{-}6)$$

The difference is taken in such a manner as to give a positive number.

The values of half-cell potentials can be calculated as a function of the activities of OX and RED by the *Nernst equation:*

$$E = E^\circ + \frac{RT}{nF} \ln \left(\frac{a_{OX}}{a_{RED}}\right)$$
$$= E^\circ + \frac{RT}{nF} \ln \left(\frac{C_{OX}}{C_{RED}}\right) + \frac{RT}{nF} \ln \left(\frac{\gamma_{OX}}{\gamma_{RED}}\right) \qquad (1\text{-}7)$$

where the γs are the activity coefficients. The constant term E°, listed for many systems in the appendix, is defined as the potential of a cell consisting of the half-cell under consideration together with a hypothetical half-cell where the reaction is $H^+ + e^- \longrightarrow \frac{1}{2} H_2$, with activities of all species equal to unity. (This reference half-cell is the *Standard Hydrogen Electrode, SHE,* and E° is called the *standard potential* of the particular redox couple.) Once E° is known, the desired half-cell potential can be calculated using Eq. (1-7) together with the appropriate activities. Concentrations are often used instead. This is a possible source of error, but in many cases the activity coefficients of OX and RED have similar values and partially compensate in the ratio.

It is more convenient to combine the activity coefficient contribution with E° in a new quantity called the *formal potential, $E^{\circ\prime}$.* With this convention, concentrations can be used directly in the Nernst equation:

$$E = E^{\circ\prime} + \frac{RT}{nF} \ln \left(\frac{C_{OX}}{C_{RED}}\right) \qquad (1\text{-}8)$$

It is important to note that $E^{\circ\prime}$ is a function of the composition of the solution including complex formation.

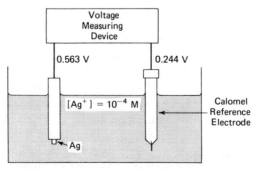

Figure 1-6. A potentiometric measurement. The voltages indicated are referred to the SHE.

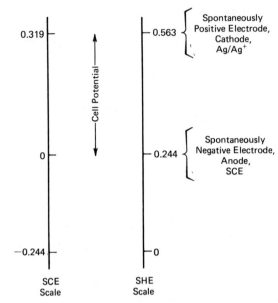

Figure 1-7. Diagram for obtaining cell potentials and polarities.

One must not forget that Eqs. (1-7) and (1-8) apply only for simple reactions such as $OX + ne^- \longrightarrow RED$; if the reaction involves other substances, such as seen in Eq. (1-2), the additional species must also be taken into account. If solids or the solvent take part in the reaction, their activities are taken as unity.

Let us consider an example of a potentiometric measurement making use of the Nernst equation, as shown in Figure 1-6. The reference electrode is a commercially available saturated calomel electrode (SCE), with a potential of about +0.244 V with respect to the SHE defined above. The potential of the silver electrode is given by $E = 0.800 + 0.0592 \log (10^{-4}) = 0.563$ V. The constant 0.0592 represents the value of $2.303(RT/F)$ at 300 K.† The potential of the cell is then 0.563 – 0.244 = 0.319 V. Because the SCE was used as a reference, this voltage is referred to as "versus SCE."

The choice as to which voltage to subtract from the other in order to obtain the electrode potential may be clarified by the diagram of Figure 1-7. The potential of the cell is represented by the distance along the scale between the two half-reactions. The higher electrode on the scale is the cathode.

When the zero of the scale is taken as the standard hydrogen electrode, the potentials are referred to as "vs. SHE." The SHE has been accepted as an international standard for the potential scale for many years, and it might appear that it has some special intrinsic value. This is not the case, however; other choices could have been made equally well. As a matter of practicality, the saturated

---

†The constants 0.0592 and 2.303 (= ln 10) are used repeatedly in this book; they are given more significant figures than usually warranted, simply for ease of recognition.

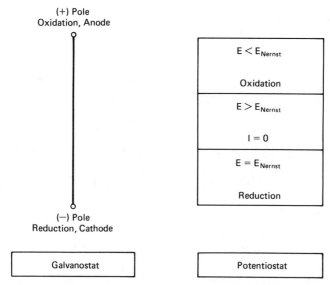

Figure 1-8. Diagrams showing the sign relations of the current and voltage. E_{Nernst} is the potential as calculated by Eq. (1-7).

calomel and Ag/AgCl reference electrodes are much more widely used than the SHE.

Similar diagrams can be made for galvanostatic and potentiostatic experiments, for example, Figure 1-8. These serve as mnemonic devices, helping one to remember at which pole of the galvanostat oxidation occurs and the direction of currents in the potentiostat.

REFERENCES

1. W. M. Latimer, *J. Am. Chem. Soc.*, **1954**, *76*, 1200.
2. E. A. Guggenheim, "Thermodynamics," 2nd ed., Wiley-Interscience, New York, **1950**, Chapter 10.

Chapter 2

ELECTROCHEMICAL MEASUREMENTS

The electrical quantities of interest in electroanalytical chemistry are voltages,[†] E, currents, I, and resistances, R. In AC techniques, we are concerned with impedances, Z, in place of resistances. In addition, a number of other quantities are occasionally needed. For instance, the surface tension of mercury γ_{Hg} has been essential in the elucidation of the structure of the metal-electrolyte interface [1]. Also noteworthy is the information that can be obtained upon performing spectrometric measurements on electrode surfaces.

In the present chapter, we will discuss the relation between the properties of the metal-solution interface and the measurements of voltage, current, and impedance (or resistance).

VOLTAGE MEASUREMENTS

The objective of a potentiometric experiment is the measurement of the difference of potential between two electrodes, E_{cell}, while no current flows. It is of some interest to consider how this can be broken down into the constituent voltages for each portion of the cell.

A question that might arise is why one always measures *differences* of potentials rather than the separate values for the two electrodes. The existence of such absolute potentials has preoccupied chemists for a long time, and it appears that we must be resigned to the fact that it is not possible to measure the potential of a single electrode [2; see, however, 3–5]. Moreover, one cannot even measure the difference of potential between two dissimilar chemical phases. The only mea-

[†]We will generally use the terms *voltage* and *potential* interchangeably. It is preferable, though, to reserve the term "potential" for the cell and to use "voltage" for external measurements. There is a third synonym, *electromotive force* (*emf*), which has fallen into disuse in this context. In this book, capital letters E and I are used for voltage or potential and current, except when it is desired to stress particular conditions.

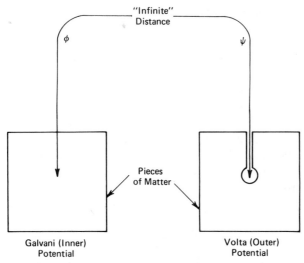

Figure 2-1. Illustrating the definitions of the Galvani and Volta potentials. It is assumed that no chemical interactions occur.

surable quantity is the difference of potential between two portions of chemically identical material. This statement appears to contradict everyday experience, which says that one can connect a voltmeter to anything he pleases. True, but in the measurement, eventually the connections will terminate at two points on the meter, made of the same metal, usually copper, and this is where the voltage is actually measured.

Before going further, let us try to define the concept of potential. The electrical potential of a phase† can be defined as the work necessary to bring a unit charge from a great distance ("infinity") to the *interior* of the phase. The charge must not alter the potential of the phase by its presence and must not enter into any chemical interaction. The potential so defined is called the *inner*, or *Galvani*, *potential*, ϕ, and is not a measurable quantity.

It is possible to define a very similar quantity, the *outer*, or *Volta*, *potential*, ψ, which is accessible to measurement. This concept can be derived from the model of the Galvani potential by removing the condition that the charge must actually penetrate the phase. It is now only required to approach it very closely. In order to avoid the effects of external electric fields, the "outside" is often considered to be the interior of a small cavity as shown in Figure 2-1, hence the alternative name of *cavity potential*. The difference between ϕ and ψ is known as the surface potential, χ. Of the three, only the outer potential ψ is measurable, although the inner potential ϕ has a clearer thermodynamic significance. Even though inner potentials cannot be measured directly, some combinations of them are accessible to experiment.

---

†The concept of "phase," as here used, is that associated with the Gibbs Phase Rule.

Consider the electrochemical cell of Figure 2-2a. In accordance with the above discussion, the difference in inner potential between Cu(A) and Cu(B) is measurable, but none of the intermediate potentials can be observed. In part b of the figure is shown one possible sequence of potential values through the cell. Note that, in an analytical context, although the cell abounds in unmeasurable potentials, the overall cell potential *is* accessible and responds to the concentrations of the various chemical species involved.

Special mention must be made of the potential difference between the two solutions A and B. This *liquid junction potential*, E_{jnct}, is present whenever two different electrolytic solutions come in contact. It is not measurable except in a restricted sense. Consider a cell with two identical electrodes immersed in dissimilar solutions. A difference of potential, E_{cell}, is exhibited that can be assigned (under certain conditions) to the junction potential. Thus [6] the cell:

$$Hg \mid Hg_2Cl_{2(s)}, HCl_{0.1M} \parallel KCl_{0.1M}, Hg_2Cl_{2(s)} \mid Hg \qquad (2\text{-}1)$$

would be expected to have almost identical activities of Cl^- ion on both sides, and therefore it can be assumed that the cell voltage actually measures the potential

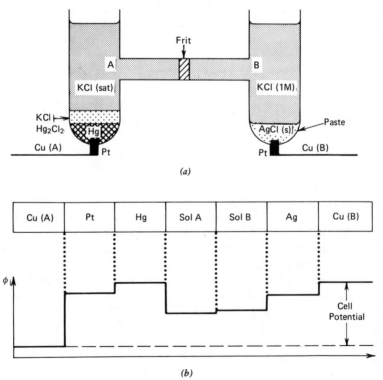

(a)

(b)

Figure 2-2. (a) An electrochemical cell and (b) some possible internal potentials. Only the overall potential can be measured.

of the liquid junction (indicated by a double vertical bar). For this particular case, the junction potential is estimated [7] to be 27.3 mV. This could constitute a significant contribution to the overall voltage of the cell and needs appropriate compensation. The liquid junction potential can be minimized by selecting as the electrolyte for the reference compartment a concentrated solution of KCl or NH_4NO_3. These particular salts generate very small junction potentials against most aqueous solutions. For example, the cell of Figure 2-2, which uses KCl, has a junction potential of about 0.5 mV which can safely be ignored for most purposes.

Voltage Measurements with Finite Current

In the previous discussion, the current passing through the cell was considered to be very small. In contrast, if appreciable current is allowed to pass, the following types of alterations in the voltage may occur:

1. An ohmic drop, usually referred to as an *IR-drop*, caused by the transport of electricity through the solution and various conductors.

2. *Concentration overpotential*, η_c, a shift in potential caused by changes in the concentrations of OX and RED in the immediate vicinity of the electrodes. The redox process causes an increase in the concentration of one species and a corresponding diminution of the other. This change is partially offset by diffusion or stirring, but the concentrations at the electrode surface, C_{ox}^s and C_{red}^s, will still be different from those in the bulk of the solution, C_{ox}^* and C_{red}^*. This causes a shift in the potential as calculated by the Nernst equation, to a new potential, E^s:

$$E^s = E^* + \eta_c = E^* + \frac{0.0592}{n} \log \left(\frac{C_{OX}^s \cdot C_{RED}^*}{C_{RED}^s \cdot C_{OX}^*} \right) \qquad (2\text{-}2)$$

where E^* is the potential exhibited when the surface concentrations are equal to the bulk values. This expression can be obtained by taking the difference between the Nernst expressions written for the two sets of concentrations. Note that, in a sense, E^s is an equilibrium quantity, since, if the solution were to be suddenly replaced by one with concentrations C_{OX}^s and C_{RED}^s extending throughout the solution, the electrode potential would remain at equilibrium, with the value E^s.

3. *Charge-transfer overpotential*, η_t, which is the excess voltage necessary to accelerate the redox reaction to the desired rate. It is a kinetic rather than a thermodynamic quantity.

In light of the above, we can write for the total voltage, E, of a cell that uses a reference electrode:

$$E_{cell} = E_{indic} - E_{ref} = \left(E^\circ + \frac{0.0592}{n} \log \frac{C_{OX}^*}{C_{RED}^*} + \eta_c + \eta_t + IR \right) - E_{ref} \qquad (2\text{-}3)$$

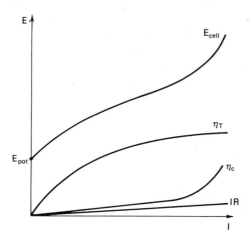

Figure 2-3. The potential E_{cell} and its various components. The detailed shapes of the curves depend on the nature and concentrations of the species present. E_{pot} is the potential that would be measured in the absence of current.

The last three terms within the brackets contain contributions from the indicator electrode (Figure 2-3); the reference electrode exhibits similar terms, but these are usually negligible.

IMPEDANCE MEASUREMENTS

Consider the curve describing the current-voltage relations in a three-electrode system, as shown in Figure 2-4. The representation can be either E versus I or I versus E. In either case, the interrelationship between the two variables is best expressed by the derivative of the curve at the particular point of interest, in this example at the coordinates E and I. Depending on the choice of coordinates, the derivative is either the dynamic impedance, $Z_{diff} = dE/dI$, or the *dynamic admittance*, $Y_{diff} = dI/dE$. Of the two, the dynamic admittance is the more interesting to the analytical chemist, since it is often proportional to the concentration of the active species. (Impedance is the more widely used term in electronics.) Note that the admittance and impedance are reciprocals of each other. Either of the two can be determined by taking the slope of the curve at any point. In practice, an alternative method of measurement in which a small AC voltage, $E_{AC} = A \sin \omega t$, is superimposed on the DC voltage, E_1, is often used (Figure 2-5).

The electrode will then exhibit simultaneously a DC current, accompanied by a small AC current, $I_{AC} = B \sin \omega t$. The admittance is measured by the ratio of the AC current and voltage:

$$Y_{diff} = \frac{I_{AC}}{E_{AC}} = \frac{B \sin \omega t}{A \sin \omega t} = \frac{B}{A} \tag{2-4}$$

where it is assumed that the phases of the two AC signals coincide, a feature that will be discussed in a later chapter. The quantities A and B are directly measurable.

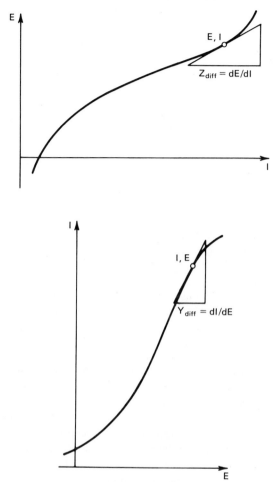

Figure 2-4.   Current–voltage characteristic of a three-electrode system, plotted in two different ways.

At this point it might be useful to clarify the relationship between the differential admittance and impedance, Y_{diff} and Z_{diff}, on the one hand, and the integral quantities, Y and Z, on the other:

$$Y = I/E \qquad Y_{diff} = dI/dE \qquad (2\text{-}5)$$

$$Z = E/I \qquad Z_{diff} = dE/dI \qquad (2\text{-}6)$$

It is important to note that the two types of definitions give identical results if the I versus E curves are straight lines passing through the origin (*linear systems*). A great number of electronic circuits show linear response, specifically all those consisting of combinations of resistors, capacitors, and inductors; hence, in electronics, the integral definition is preferred as being simpler.

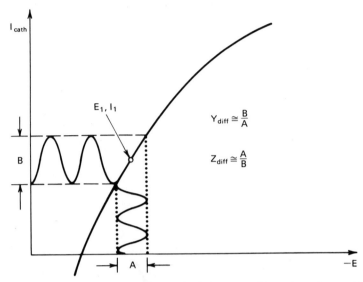

Figure 2-5. Determination of the differential admittance and impedance by an AC method. In practice the AC amplitude is made quite small, in order to enhance resolution.

Electrochemical systems are almost never linear, because of the presence of *faradaic impedance*, Z_{far}, which is a function of voltage, representing the contribution of the redox process to the actual shape of the I versus E curve. (Correspondingly, we can speak of the *faradaic admittance*.) It is because of this effect that it is preferable to use differential quantities. The integral definitions would require knowledge of the potential with respect to some reference (Figure 2-6); any change in the reference would change the value of Z, a complicating factor. The qualifier "differential" will be considered implicit in the rest of this book.

In addition to the faradaic impedance, the electrode system contains contributions due to ohmic resistance and to capacitance. A typical electrode model, expressed in terms of impedances, is shown in Figure 2-7. The component R represents the resistance of the solution and of the connecting wires. The notation Z_{dl} refers to the impedance of the capacitor C_{dl} formed by the electrical charges on each side of the electrode-solution interface, the so-called *electrical double layer*. The relationship between the impedance and the capacitance of the double layer is given by the expression:

$$Z_{dl} = \frac{1}{2\pi f C_{dl}} \tag{2-7}$$

where f is the frequency in hertz and C_{dl} is measured in farads. The double-layer capacitance is defined differentially in line with the definition of impedance. It represents the increase in charge, dQ, that results when the potential is increased

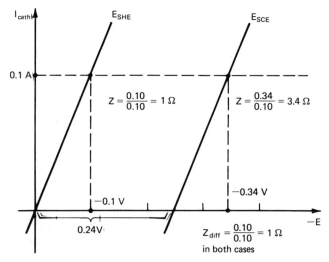

Figure 2-6.   The effect upon Z and Z_{diff} caused by changing the reference point for a given electrode.   The plots are shown as straight lines for simplicity, but may be curved.   The differential impedance is unchanged by the shift in reference, since the slopes of both lines are the same.

by dE:

$$C_{\text{dl}} = \frac{dQ}{dE} \tag{2-8}$$

It is important to distinguish between the capacitance (in farads), which is frequency-independent, and the capacitive impedance (in ohms), which changes with frequency.

It is instructive to return now to Figure 2-7 and to discuss various limiting cases. Consider first DC conditions; there the impedance of the capacitor can be taken as

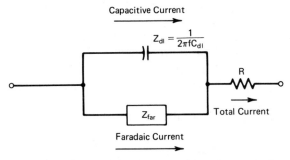

Figure 2-7.   Representation of an electrode in terms of impedance.   Such a representation is called an equivalent circuit.

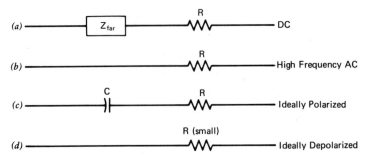

Figure 2-8.   Equivalent circuits for electrodes under various limiting conditions.

infinite, since no current can pass through it. Hence the capacitance can be eliminated from the equivalent circuit, which reduces to that shown in Figure 2-8a. If,
in contrast, the frequency is relatively high (say 10 kHz), the value of Z_{dl} becomes
small, and practically all the current passes through the double-layer capacitance.
The contribution of the faradaic impedance is then negligible, and the only element
affecting the current is the resistance of the solution, as shown in Figure 2-8b.
This represents the case sought in conductometric measurements.

A third situation (Figure 2-8c) arises if there are no active redox processes
present, so that Z_{far} can be eliminated from the circuit. In this case, at DC, the
electrode acts as an open circuit and no current flows. At AC, all the current
passes through the double-layer capacitance. Such an electrode, where no chemical
work is done, is said to be *ideally polarized* and is useful for investigating the
properties of the double layer. Mercury in contact with an inert electrolyte forms
a good approximation to an ideally polarized electrode and has played an important
role in the development of electrochemistry [1].

Finally, if the faradaic impedance is zero, the capacitor will be effectively shorted
out, and again the resistance of the solution will remain in control of the current,
as seen in Figure 2-8d. If, in addition, the IR-drop is small, the electrode is called
ideally depolarized. This type of electrode exhibits a voltage that is independent
of the current, and hence is useful as a reference electrode. The qualifier "ideal"
should not be taken too seriously, since all electrodes exhibit some resistance and
faradaic impedance.

In Figure 2-9, the I versus E curves corresponding to the two "ideal" cases are
presented. It can be ascertained that these curves are characterized by a differential
impedance equal to zero and infinity, respectively.

THE ELECTRICAL DOUBLE LAYER

The existence of accumulations of charge on both sides of the electrode-solution
interface has been recognized for a long time and is described as an electrical double
layer. More recent investigations have shown it to be more complex in structure
than first thought, but the term continues to be used, even though it is not fully

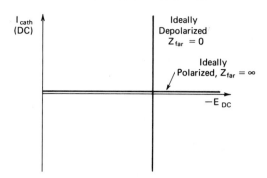

Figure 2-9.   Current–voltage curves for ideally polarized and depolarized electrodes. The curve for the polarized electrode is actually coincident with the voltage axis.

descriptive of the actual structure of the interface. The present picture is fundamentally the model proposed by Stern in 1924, with a few minor modifications [1, 8, 9].

Let us assume that the metallic electrode has an excess of charges q_m of either positive or negative sign. To compensate for it electrically, three ionic zones will form in the solution (Figure 2-10). One layer of ions, actually touching the electrode surface, defines the "plane of closest approach" passing through the centers of these ions, called the *inner Helmholtz plane, IHP.*

The next layer of ions is defined as the *outer Helmholtz plane, OHP.* This is where most cations and some anions are encountered. Finally there is a diffuse layer of mixed charges that extends into the body of the solution. The sum of all the charges in the three solution layers, q_i, must be equal and of opposite sign to the charge on the metal, q_m. The IHP and OHP together constitute the *compact layer* of charges. It is very strongly held by the electrode and can survive even when the electrode is pulled out of the solution [10].

The IHP region contains mostly molecules of solvent. Some ions, especially if not strongly solvated, and under appropriate potential conditions, can displace solvent molecules and enter the IHP as *specifically adsorbed ions.* They are in most cases anions, but can sometimes be cations or even ion-pairs [11]. The forces retaining them are dependent on the nature of the ion as well as on the potential. In contrast, the ions in the next layer, the OHP, conserve their solvation spheres and interact electrostatically with the other charged species and with the field of the electrode.

The capacitance of the double layer consists of the combination of the capacitance of the compact layer in series with that of the diffuse layer [1]. Of the two, the diffuse layer is dependent on the amount of solute, and changes dramatically in thickness with the concentration. For example, in a 0.1 M solution, it might extend only to about 10^{-5} mm from the electrode, while in a dilute solution it might reach as far as 10^{-3} mm. In contrast, the compact layer is perhaps only 5×10^{-7} mm thick and largely independent of the concentration. In addition to its effect on the capacitance, the variation in the thickness affects also the drop

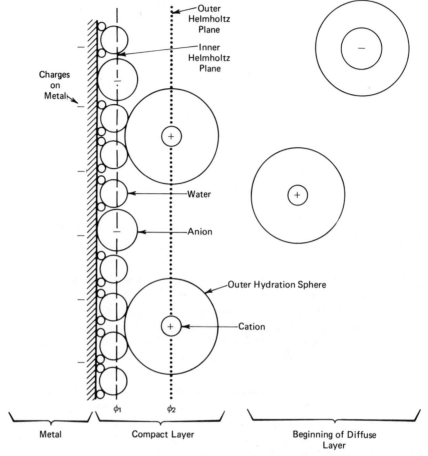

Figure 2-10. The electrical double layer in aqueous solution. The spaces between ions are assumed to be occupied by water molecules, a few of which are shown. The charge configuration varies with the potential of the electrode.

of potential across the diffuse layer.† To avoid the complications due to these effects, a *supporting electrolyte* of concentration at least 0.1 *M* is customarily employed; this is an inert electrolyte that contributes to the transport of electricity by the solution but is faradaically inactive.

Double-layer capacitances can be satisfactorily measured with rather simple devices, such as seen in the exquisite work of Grahame [12], but modern, sophisticated measurements [13] are considerably more convenient.

The measured differential capacitance exhibits a rather strong dependence on the voltage and on the nature of the electrolyte, especially on the anion. A few typical examples are shown in Figure 2-11. The sharp dip occurs only with dilute solutions

†Often referred to in the literature as ϕ_2.

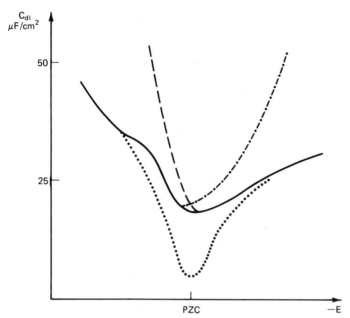

Figure 2-11. Typical differential capacitance of the mercury–water interface. Full line: no adsorption, $10^{-1}\,M$ electrolyte. Dotted line: same, but $10^{-3}\,M$ electrolyte. Dashed line-specifically adsorbed anion. Dash-dot line: specifically adsorbed cation. Large variations exist in specific cases. The meaning of PZC will be explained in the next section.

and is a contribution of the diffuse layer. The rest of the curve is determined mostly by the compact layer.

ELECTROCAPILLARITY

The study of the interfacial tension γ between the metal (usually mercury) and a solution of electrolyte as a function of potential is called *electrocapillarity*.† Superficially the surface tension might appear alien to electrochemistry; consider, though, that two fundamental features are common to both: an electrical potential, and a mercury-solution interface. Because of this similarity, electrocapillarity is very useful in elucidating the properties and structure of the double layer. In fact, the charge on the electrode, Q, can be calculated from purely electrocapillary data [1]:

$$Q = -\frac{\partial \gamma}{\partial E} \tag{2-9}$$

†The name is derived from an instrument, the Lippmann capillary electrometer, that was used in the early stages of investigation in this area.

It is possible to go one step further, recalling that $C = dQ/dE$, and write:

$$C_{dl} = -\frac{\partial^2\gamma}{\partial E^2}$$

(2-10)

Note, however, that the precision of obtaining second derivatives is, as a rule, rather poor.

Electrocapillary data are usually represented as graphs of γ versus E, as shown in Figure 2-12. A definite potential dependence is obtained, which is due to the repellent effect of the charges accumulated at the electrode surface. The maximum, called the *electrocapillary maximum* (*ECM*), nearly coincides with the *point of zero charge* (*PZC*) for the double layer, which is the point at which the sign of the electrode charge reverses as the result of the applied potential. Charges of either sign produce the same repulsion, and the curve might be expected to be symmetrical about the PZC. In practice, the presence of specific adsorption at the inner Helmholtz plane causes deviations from symmetry. Neutral molecules are often adsorbed as well, radically modifying the electrocapillary curve and the differential capacitance.

The PZC is in principle a valid reference point, and a potential referred to it is called a *rational* potential. This choice has the serious disadvantage that the point varies with the nature and concentration of solutes present, and thus finds little analytical use.

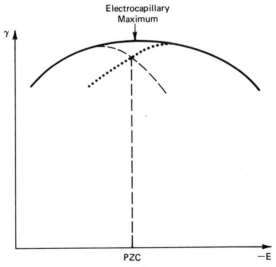

Figure 2-12. The surface tension of mercury in aqueous solutions as a function of the applied potential. Dashed line: specifically adsorbed cation. Dotted line: specifically adsorbed anoin, the more common case. The surface tension decreases some 25 percent on either side of the maximum, which for mercury in a nonadsorbing electrolyte is about -0.5 V (vs. SCE).

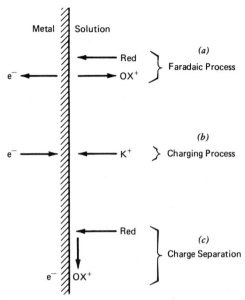

Figure 2-13. Fundamental types of electrode processes. In this example the faradaic reaction is written for simplicity as an oxidation: $RED = OX^+ + e^-$. It is assumed that the K^+ ion is not reduceable at the applied potential.

CURRENT MEASUREMENTS

The current† is a measure of the rate of an electrochemical reaction, but there are some caveats. One restriction results from the fact that electrochemical reactions are frequently associated with interfering chemical processes not directly controlled by the current. In this case, the overall extent of the reaction may not be immediately predictable from current measurements.

In addition, the current may have components that are not due to the redox process. Such *nonfaradaic* currents are primarily caused by alterations in the double-layer charge; they are called *charging* or *capacitive currents*. The correction for charging currents is often a difficult matter.

In Figure 2-13 the fundamental types of electrode processes are shown schematically in terms of particle transport. In type (*a*), the faradaic process, the species RED is shown as being transported to the metal electrode where it loses an electron; it then moves away as OX^+. There is a net current resulting from the simultaneous movement of positive and negative charges in opposite directions away from the interface. In contrast, (*b*) represents a charging process where no redox reaction takes place. The current is due to the simultaneous motion of

†It must be noted that the symbol I is sometimes used for *current density* rather than for current itself. This we consider poor practice, and in this book current density will always be denoted by I/A.

charges towards, but not across, the interface. Evidently, this process cannot continue indefinitely, but only until the voltage reaches its final value. Because of this, the charging current is a *transient* phenomenon, that is, a process that occurs for a short time only, after each change in conditions.

The process shown at (c) is also a transient [14, 15]. It consists of a local dissociation of the RED $\longrightarrow OX^+ + e^-$. This is, for all practical purposes, an oxidation, but it is not accompanied by current, since there is no complete charge separation, but only the formation of a dipole with OX^+ and e^- facing each other across the interface. This process can be troublesome, since it contributes to the double-layer charge without contributing to the charging current; it is minimized by the presence of a supporting electrolyte.

The charging process can be evaluated if we note that the total charge on an electrode can be expressed as:

$$Q = C \cdot A \cdot E \tag{2-11}$$

where C is the integral capacitance per unit area, A is the area of the electrode, and E is the potential referred to the PZC.

Note that the current is the time derivative of the charge, dQ/dt, and therefore,

$$I_{chg} = \frac{dQ}{dt} = CA\frac{\partial E}{\partial t} + CE\frac{\partial A}{\partial t} + AE\frac{\partial C}{\partial t} \tag{2-12}$$

The first term on the right represents the conventional charging of the capacitor as the voltage increases. The term in $\partial A/\partial t$ is of importance in polarography, where the area of the electrode varies widely although the voltage changes only slowly. The last term in the equation is important when processes, such as adsorption, change the double-layer capacitance.

In light of the discussion above, one can write:

$$I_{far} = I - I_{chg} \tag{2-13}$$

or

$$\text{Number of moles} = \frac{I_{far}}{nF} \cdot t = \frac{I - I_{chg}}{nF} \cdot t \tag{2-14}$$

which permits the calculation of the extent of the reaction during the time period t. The factor nF is the conversion factor between chemical and electrical quantities. Correction for the charging current is not always feasible and in such cases, the ultimate sensitivity of the analytical method may be compromised.

The faradaic current itself is related to three consecutive processes: (1) mass transport to the electrode; (2) reduction or oxidation, and (3) mass transport of the products away from the electrode surface. In some cases, additional effects exist, such as chemical reactions, adsorption, or formation of solids.

Consider the case where such complications are absent, and moreover, assume that the solution is unstirred. The overall process can be described as:

$$(OX)^* \xrightarrow{r_{ox}} (OX)^s \xrightarrow{r_t} (RED)^s \xrightarrow{r_{red}} (RED)^* \qquad (2\text{-}15)$$

where the symbols $*$ and s refer to the bulk and electrode surface situations, respectively. The rates r_{ox} and r_{red} are the corresponding mass transport rates (in this instance by diffusion), and r_t is an electron transfer rate. The units for all rates are mol $\cdot$ cm^{-2} $\cdot$ s^{-1}.

The rates of diffusion are functions of time and concentrations, and reflect the properties of the species involved through a constant called the *diffusion coefficient*, D, which carries the units cm^2 $\cdot$ s^{-1}. In general, D_{ox} and D_{red} are similar in value, and of the order of 10^{-5}.

By contrast, the electron-transfer *rate* varies widely from system to system. It also depends on the concentration of the reagent and on the potential applied.

Upon examining Eq. (2-15), it can be seen that in accordance with conventional chemical kinetics the slowest process will determine the overall rate. If we assume first-order kinetics for all steps, then the various rates will depend on concentrations in a similar way, and it follows that an interplay of potential E and time t will define which of the rates are slow and thus dominate the process. Noting also, as mentioned above, that $D_{ox} \cong D_{red}$, we can assume that r_{ox} and r_{red} are not too different, so that only three cases are likely to occur:

1. Both rates r_{ox} and r_{red}, being nearly equal and small, dominate the kinetics. This process is time-dependent, and is called *reversible*, or more exactly electrochemically reversible.

2. The electron-transfer rate, r_t, is dominant. In this case the process is time-independent. This is referred to as electrochemically *irreversible*.

3. All rates are comparable in value, so that the kinetics is mixed and rather complicated. This case is referred to as *quasi-reversible*.†

Let us consider a case where the potential applied to an electrode is the same as the equilibrium potential E^s as calculated from surface concentrations by the Nernst equation, and assume that E^s and E^*, as defined in Eq. (2-2) are not very different. Since the system is in equilibrium, no current will pass, and $r_t = 0$. The kinetic control belongs to the charge transfer which keeps the process stationary. If a small deviation from E^s is introduced, the rate will become non-zero but will remain small. The process continues to be charge-transfer controlled, and we can write for the reaction:

$$OX + ne^- \underset{r_b}{\overset{r_f}{\rightleftharpoons}} RED \qquad (2\text{-}16)$$

---

†Other meanings of the term "quasi-reversible" are sometimes encountered in the literature.

The rate equation is:

$$r_t = r_f - r_b = k_f C_{OX}^s - k_b C_{RED}^s \tag{2-17}$$

in which *two* rates, forward and backward, are present simultaneously. The rate constants k_f and k_b correspond to the processes of reduction and oxidation, respectively. Both rate constants are exponential functions of potential and of a parameter called the *transfer coefficient*, denoted by α which is usually close to 0.5. It can be shown that a single rate constant, k°, will serve to replace the other two, and we can write:

$$r_t = k^\circ C_{OX}^s \exp\left\{-\frac{\alpha nF}{RT}(E - E^{\circ\prime})\right\} - k^\circ C_{RED}^s \exp\left\{\frac{(1-\alpha)nF}{RT}(E - E^{\circ\prime})\right\} \tag{2-18}$$

where E represents the applied potential.

This equation is fundamental to electrochemical kinetics, and was developed through the efforts of Erdey-Gruz and Volmer, and of Butler. If the current rather than the rate is desired, the expression must be multiplied by nFA, where A is the area of the electrode. The *standard heterogeneous rate constant*, k°, corresponds to the common value of k_f and k_b at the formal potential, $E^{\circ\prime}$, as can be seen by inspection of Eqs. (2-17) and (2-18).

An alternative form of the equation can be obtained if the potentials are referred to E^s, the equilibrium potential corresponding to the surface concentrations. In this case, the argument of the exponentials will contain the charge-transfer overpotential η_t, defined as $E - E^s$, and the expression takes the form:†

---

†If we note the identity: $\exp(a) = \exp(a - b)\exp(b)$, it is seen to be possible to shift the potential from one reference to another by writing Eq. (2-18) in the form:

$$I = nFAk^\circ \left(C_{OX}^s \exp\left\{-\frac{\alpha nF}{RT}(E^s - E^{\circ\prime})\right\}\exp\left\{-\frac{\alpha nF}{RT}(E - E^s)\right\}\right.$$

$$-\, C_{RED}^s \exp\left\{\frac{(1-\alpha)nF}{RT}(E^s - E^{\circ\prime})\right\}$$

$$\left. \cdot \exp\left\{\frac{(1-\alpha)nF}{RT}(E - E^s)\right\}\right) \tag{2-18a}$$

Note that $E - E^s = \eta_t$, and also that the Nernst equation can be used to obtain the values of the following exponentials:

$$\exp\left\{\frac{\alpha nF}{RT}(E^s - E^{\circ\prime})\right\} = \left(\frac{C_{OX}^s}{C_{RED}^s}\right)^{-\alpha} \tag{2-18b}$$

$$\exp\left\{\frac{(1-\alpha)nF}{RT}(E^s - E^{\circ\prime})\right\} = \left(\frac{C_{OX}^s}{C_{RED}^s}\right)^{1-\alpha} \tag{2-18c}$$

When these are introduced into Eq. (2-18), they lead to Eq. (2-19).

$$I = nFAk°(C^s_{OX})^{1-\alpha}(C^s_{RED})^{\alpha}\left(\exp\left\{-\frac{\alpha nF}{RT}\eta_t\right\} - \exp\left\{\frac{(1-\alpha)nF}{RT}\eta_t\right\}\right) \quad (2\text{-}19)$$

or:

$$I = I_0\left(\exp\left\{-\frac{\alpha nF}{RT}\eta_t\right\} - \exp\left\{\frac{(1-\alpha)nF}{RT}\eta_t\right\}\right) \quad (2\text{-}20)$$

where I_0 is the exchange current given by:

$$I_0 = nFAk°(C^s_{OX})^{1-\alpha}(C^s_{RED})^{\alpha} \quad (2\text{-}21)$$

Note that in Eq. (2-20), at $\eta_t = 0$, the exponentials vanish and the total current is zero, so that I_0 describes the equal anodic and cathodic currents present at equilibrium. This is illustrated in Figure 2-14, where the plots of the two terms in Eq.

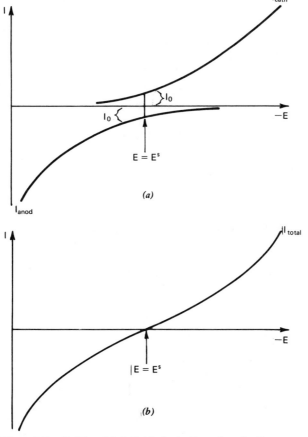

Figure 2-14. Plot of Eq. (2-20). (a) Individual anodic and cathodic currents; (b) over-all current as a function of η.

(2-20) are given both separately and as their sum. It can be seen that at $E = E^s$ the net current is zero, but there is still activity at the electrode, expressed by I_0. The exchange current definition in Eq. (2-21) contains the heterogeneous rate constant k°, and measurements can be made that permit its determination therefrom. A frequently used procedure is to plot $\log (I/A)$ versus E, as in Figure 2-15. The resulting graph, called a *Tafel plot*, has two branches separated by a negative infinity point corresponding to zero current at E^s.

A few words are in order about the charge-transfer overvoltage, η_t, that appears in these equations. The total applied potential E in a potentiostatic system is given by Eq. (2-3):

$$E = E^\circ - E_{ref} + \frac{RT}{nF} \ln\left(\frac{C_{OX}^*}{C_{RED}^*}\right) + \eta_c + \eta_t + IR \qquad (2\text{-}22)$$

from which

$$E = E^s + \eta_t + IR \qquad (2\text{-}23)$$

It can be seen that E^s contains all the potentials except the charge-transfer overvoltage and the ohmic drop. A representation of the relations between the various voltages is given in Figure 2-16.

The origin of the voltage axis in Figure 2-16 is the standard electrode potential, E°. The voltage marked $E^\circ - E_{ref}$ is the standard potential of the redox couple with respect to the reference. The point marked E^* is the voltage that would be measured potentiometrically (zero current) between the indicator and reference electrodes. The overvoltages are measured from this point.

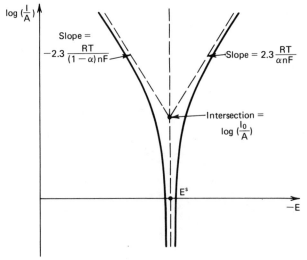

Figure 2-15. A Tafel plot.

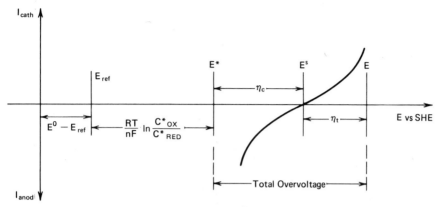

Figure 2-16.   Graphical representation of Eq. (2-22). The values of $(E^{\circ} - E_{ref})$ and E^* could be either positive or negative. It is assumed that the IR product is negligible.

The point marked E^s indicates the potential that one would measure potentiometrically with the concentrations at the electrodes as altered by the passage of current rather than equal to the bulk concentrations. Finally, the unsuperscripted E is the voltage applied to the indicator electrode by a potentiostatic or galvanostatic circuit.

Of special interest is the faradaic current flowing when the overvoltage η is very small. In this case, the exponential terms involved in the rate equation can be expanded by the approximation: $\exp(x) = 1 + x$. If the expansion is performed on Eq. (2-20), we obtain:

$$I = -I_0 \frac{nF}{RT} \eta_t \tag{2-24}$$

which gives to the quantity $I_0 nF/RT$ the dimensions of an admittance, the reciprocal of the faradaic impedance. Equation (2-24) can be very useful in determining kinetic parameters.

In practice, the major factor that influences the relative magnitudes of the various potentials indicated is the value of I_0 for the particular redox couple present at the specified concentrations. If the exchange current is large, Figure 2-17a shows that, for a given current, the charge-transfer overvoltage η_t must be small. In this case, the kinetic behavior is governed by the movement of ions to and from the electrode (*transport control*), and the electrode is reversible. If diffusion is the means of transport, the process is said to be under *diffusion control.*

The opposite case, where the exchange current is small, is shown in Figure 2-17b. The current remains low over a wide range of voltages, and the overvoltage is substantial, even if only small currents are called for. Kinetic control now belongs to the charge-transfer process. It is said to be under *charge-transfer control*, and the process is irreversible. Note that, as mentioned before, the time also enters into the kinetic equations. As a general rule, if the duration of an experiment is made short, the process tends to become less reversible, and vice versa.

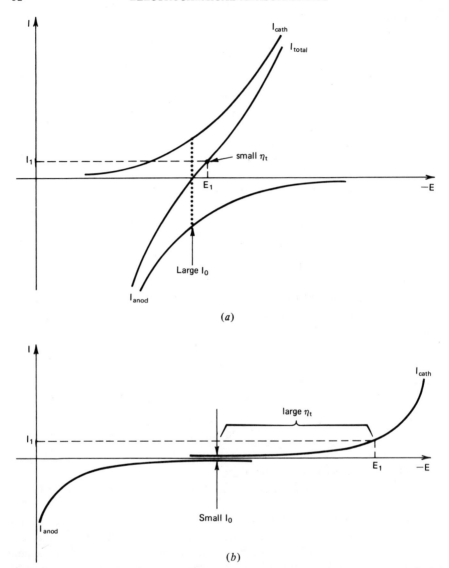

Figure 2-17.   The effect of the exchange current on the overvoltage.   The electrode in (a) is depolarized, that in (b) is polarized.

The concept of reversibility is used frequently to qualify current-voltage curves, since the ideal shape occurs only for the reversible case. Irreversible systems show somewhat ill-defined curves. The choice of these terms for the present purpose is rather unfortunate, since they do not coincide with the thermodynamic concepts. Thermodynamic reversibility is defined in terms of the possibility of back-tracking any change in the system by exactly the same path. A concise definition is, "A process is thermodynamically reversible if it is described by relations that are in-

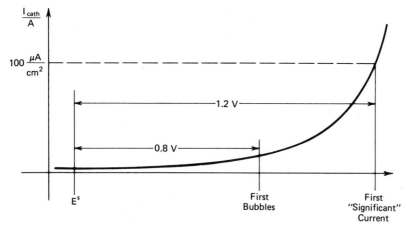

Figure 2-18.   Two different values for the "overvoltage" of hydrogen on mercury. Similar curves could be drawn for other irreversible systems. The values 0.8 and 1.2 V are illustrative only.

variant with respect to the sign affixed to the variable *time*." This condition is almost never fulfilled for electrochemical cells, if current is passing.

Finally, we would like to mention another use of the term overvoltage, or overpotential, in relation to the evolution of hydrogen at a cathode. This represents the voltage span between the equilibrium potential and the point where a significant amount of current flows (see Figure 2-18). The difficulty with this concept lies in the interpretation of the word 'significant.' In the example shown in the figure, if 'significant' means that the first bubbles of hydrogen are observed, then the overvoltage is about 0.8 V. However, if "significant" refers to some arbitrary current density, say 100 $\mu A/cm^2$, then the overvoltage becomes 1.2 V.

DIFFUSION TRANSPORT

The redox process at the electrode in a quiet solution will deplete one of the constituents, say OX, and create an overpopulation of RED. Consequently, diffusion must serve to transport the species OX toward, and RED away from, the electrode. The driving force for diffusion is the difference between the bulk and surface concentrations. The situation is illustrated in Figure 2-19. As time increases, the concentration profiles become less steep. This affects the rate of diffusion, since by Fick's law, the flux of matter, Φ, is directly proportional to the concentration gradient:

$$\Phi = -AD_{OX}\left(\frac{\partial C_{OX}}{\partial x}\right) \qquad (2\text{-}25)$$

This is written for OX, but a similar relation holds for RED. In this equation, D_{OX} is the diffusion coefficient.

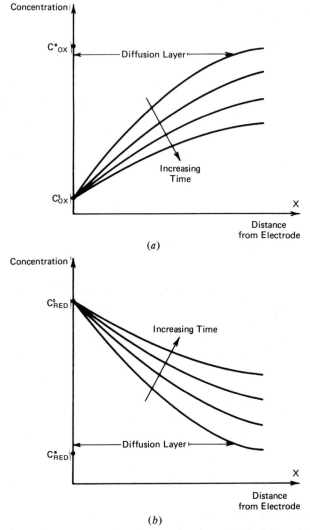

Figure 2-19.   Concentration profiles at an electrode. The depth of the diffusion layer is of the order of $\sqrt{Dt}$.

If we take into account the dependence of the flux on the gradient, we observe that it must decrease with time. It can be shown that the actual time dependence of the flux toward the surface, Φ^s, is:

$$\Phi^s = A(C_{OX}^* - C_{OX}^s)\left(\frac{D_{OX}}{\pi t}\right)^{1/2} \tag{2-26}$$

We will make two modifications to this equation: (1) we will consider that the charge-transfer reaction is so fast that C_{ox}^s is negligible, and (2) we will multiply by

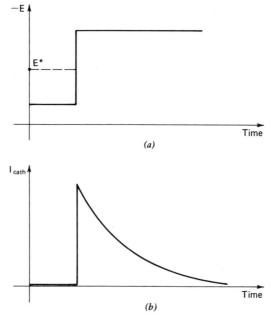

Figure 2-20. (*a*) A potential step, and (*b*) the response of an electrode in an unstirred solution.

the ubiquitous nF to obtain the current:

$$I = nF\Phi^s = nFA \left(\frac{D_{OX}}{\pi t}\right)^{1/2} C_{OX}^* \tag{2-27}$$

This is the *Cottrell equation*, an important relation fundamental to many electrochemical techniques. It represents the effect of a voltage step occuring at time $t = 0$, going suddenly from a voltage positive with respect to E^* to a much more negative voltage, as illustrated in Figure 2-20. The decrease of current follows a $1/\sqrt{t}$ law, the basic time dependence of unstirred solutions. The measurement of such currents is called *chronoamperometry*.

REFERENCES

1. D. C. Grahame, *Chem. Rev.*, **1947**, *41*, 441.
2. E. A. Guggenheim, *J. Phys. Chem.*, **1929**, *33*, 842.
3. V. D. Parker, *J. Am. Chem. Soc.*, **1976**, *98*, 98.
4. S. Trasatti, *J. Chem. Phys.*, **1978**, *69*, 2938.
5. W. N. Hansen and D. M. Kolb, *J. Electroanal. Chem.*, **1979**, *100*, 493.
6. J. J. Lingane, "Electroanalytical Chemistry," 2nd ed., Wiley-Interscience, New York, **1958**, p. 59.

7. W. H. Smyrl and J. Newman, *J. Phys. Chem.*, **1968**, *72*, 4660.
8. J. O'M. Bockris, M. A. V. Devanathan and K. Müller, *Proc. Roy. Soc. (1)*, **1963**, *A274*, 55.
9. R. Parsons, *J. Electrochem. Soc.*, **1980**, *127*, 176C.
10. W. N. Hansen, C. L. Wang and T. W. Humpherys, *J. Electroanal. Chem.*, **1978**, *93*, 87.
11. F. M. Kimmerle and H. Menard, *Can. J. Chem.*, **1977**, *55*, 3312.
12. D. C. Grahame, *J. Am. Chem. Soc.*, **1949**, *71*, 2975.
13. P. F. Seling and R. deLevie, *Anal. Chem.*, **1980**, *52*, 1506.
14. P. Delahay, *J. Phys. Chem.*, **1966**, *70*, 2373.
15. K. Holub, G. Tessari and P. Delahay, *J. Phys. Chem.*, **1967**, *71*, 2612.

Chapter 3

POTENTIOMETRY

The objective in a potentiometric measurement is to obtain information about the composition of a solution through the potential appearing between two electrodes. The measurement of the cell potential must be determined under thermodynamically reversible conditions, implying that a sufficient time must be allowed for the cell to equilibrate, and that only negligible current may be drawn during the determination.

The actual measurement is uncomplicated and can be made accurate to within a millivolt or so without too much difficulty. Either a potentiometer or an electronic voltage follower can be used.

ELECTRODE POTENTIALS

The potential of a cell can be expressed by the relation:

$$E = (E_{ind} - E_{ref}) + E_{jnct} \qquad (3\text{-}1)$$

The potential of the indicator electrode, E_{ind}, responds to the chemistry of the solution, whereas the reference has a fixed potential, E_{ref}, independent of the solution in the cell. The potential also includes the *liquid junction potential, E_{jnct}*, that appears at the interface between the electrolyte inside the reference electrode compartment and the external solution.

As was seen in Chapter 2, there is a fundamental difference between the experimental potential E and its major components, E_{ind} and E_{ref}, in that the latter quantities are not individually determinable. This situation results from the fact that, given any number N of electrodes, one can make only $N - 1$ *independent* measurements with them. Consequently, the potentials of the individual electrodes can be determined only relative to each other. It is necessary to give an arbitrary value to the potential of just *one* electrode, in order to be able to assign values to all the others [1-4]. The electrode universally accepted as the primary reference is the standard hydrogen electrode, SHE, mentioned previously. The SHE is arbitrarily given a potential of exactly zero at all temperatures. The reaction involved is:

$$H^+_{(aq,\ a=1)} + e^- \longrightarrow \tfrac{1}{2} H_{2(aq,\ a=1)} \qquad (3\text{-}2)$$

where H^+ stands for the hydrated proton. Both hydrogen gas and the proton are in contact with a metal such as platinum, that ensures reversibility of the reaction.

The choice of the SHE as the primary standard suffers from the disadvantage that it involves activities, and there is no direct way of implementing it. Precision measurements with the SHE are made by a process of extrapolation to unit activity from dilute solutions, where the effect of the activity coefficients disappears [5, 6].

Physically, the hydrogen electrode comes in many different forms, some very elaborate [7]. One of the best designs for analytical applications is the classical one by Hildebrand [8], (Figure 3-1), which can be inserted directly into a solution. The potential of the hydrogen electrode follows the relation:

$$E_{H_2} = E^\circ + \frac{RT}{F} \ln\left(\frac{a_{H_2}}{\sqrt{p}}\right) \qquad (3\text{-}3)$$

where p is the partial pressure of the hydrogen gas. The term E° is zero by definition of the SHE. The hydrogen pressure is equal to atmospheric pressure less small corrections for the hydrostatic head of a centimeter or two of water and for the presence of water vapor in the bubbles of gas. The hydrogen gas activity can usually be considered unity, if an accuracy of ± 1 mV is acceptable. Note that at unit pressure the electrode responds to the pH and, in fact, constitutes a viable pH electrode. (The error margin of 1 mV corresponds to about 0.02 pH unit.)

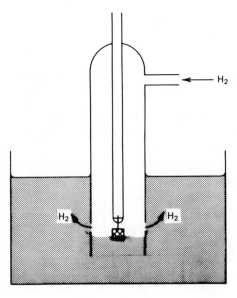

Figure 3-1. The Hildebrand hydrogen electrode.

TABLE 3-1

Selected Reference Electrodes and Their Potentials

(vs. SHE)

Electrode	Potential	
	$20°C$	$24°C$
Hg/Hg_2Cl_2, $KCl_{sat'd}$	0.248	0.244
Hg/Hg_2Cl_2, KCl_{1m}	0.282	0.280
Hg/Hg_2Cl_2, $KCl_{0.1m}$	0.334	0.334
$Ag/AgCl$, $KCl_{sat'd}$	0.204	0.199
Hg/Hg_2SO_4, $K_2SO_{4sat'd}$	-0.573	-0.577

In practice, other secondary standards are invariably used as a reference in place of the rather inconvenient SHE. Reference electrodes are defined in terms of concentrations rather than of activities, and recipes exist for their implementation [7, 9]. A few important examples are given in Table 3-1. The most common are the saturated calomel electrode (SCE) and the $Ag/AgCl$ electrode, shown in Figure 3-2. If chloride ions are objectionable, other electrodes, such as the Hg/Hg_2SO_4, can be used to advantage.

In many nonaqueous solvents, a good reference electrode is the $Ag/AgNO_3$ (0.01 M), since silver nitrate is sparingly soluble in such solvents. The hydrogen electrode can also be used in many cases, and even an aqueous calomel or $Ag/AgCl$ separated from the solution by a salt bridge containing tetraethylammonium perchlorate can be useful.

Liquid Junction Potentials

A junction potential appears whenever two phases come in contact. If the two phases are electrolytic solutions of different compositions, the potential is called a liquid junction potential, as mentioned in Chapter 2. For example, at the tip of the reference electrodes of Figure 3-2, there will be a potential difference between the saturated KCl on the inside of the electrode assembly and the solution on the outside.

The magnitude of this potential depends primarily on the gross composition of the solutions involved, so that in the presence of a supporting electrolyte there will be no appreciable changes in the liquid junction potential as the concentration of the much more dilute electrochemically active species is changed. Consequently, in methods using calibration by comparison with standard solutions, and in titrations, there normally will be little or no error from this source. Junction potentials are particularly small if one of the two solutions contains a concentrated electrolyte with nearly equal ion mobilities, such as KCl or NH_4NO_3. The inner solutions employed in most reference electrodes are of this type.

An estimate of the junction potential can be made by consideration of the very

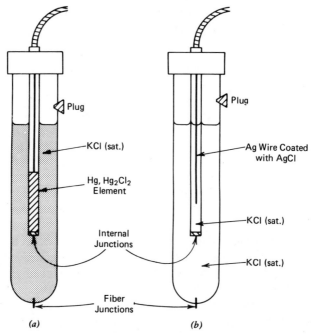

Figure 3-2. Examples of reference electrodes. (a) SCE; (b) Ag/AgCl. The plug should be removed during measurement to allow a small outward flow of solution (typically less than 20 μL/hr) to avoid contamination. The resistance of either type is around 1 kΩ.

small, but finite, current across the interface between the two solutions (Figure 3-3). The current will be transported by K^+ and Cl^- ions in proportion to their transference numbers t_+ and t_-, respectively. As each ion is transported to an area of different concentration, its chemical potential μ will change. For a small displacement, this change is $t_+ d\mu_+$ and $t_- d\mu_-$. By integration of these contrubutions across the junction, one can obtain the junction potential. The integration depends on the actual details of the transition between the two solutions, and much effort has been spent in consideration of various cases [10–12]. A few approximate values for junction potentials [9] involving saturated KCl are shown in Table 3-2. It is evident that, except for concentrated acid or base, the potentials are small and not too different from each other. When using a reference electrode containing concentrated KCl or NH_4NO_3, it is probably better to ignore the junction

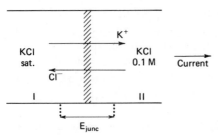

Figure 3-3. A simple liquid junction and its junction potential.

TABLE 3-2
Estimated Liquid-Junction
Potentials of the Type:
Solution||KCl$_{sat'd}$

Solution	$E_{jnct}(mV)^a$
HCl, 1 M	14.1
HCl, 0.1 M	4.6
HCl, 0.01 M	3.0
KCl, 0.1 M	1.8
KH Phthalate, 0.05 M	2.6
NaOH, 0.01 M	2.3
NaOH, 1 M	-8.6

^{a}Note: In order to obtain a corrected potential, the number in the table should be added to the observed potential of the indicator electrode.

potential altogether, since there is great doubt about the validity of various corrections.

The Indicator Electrode

In Chapter 1, it was shown that the potential of an indicator electrode obeys the Nernst equation, which can be written in the form:

$$E = E^\circ + \frac{RT}{nF} \ln\left(\frac{\gamma_{OX}}{\gamma_{RED}}\right) + \frac{RT}{nF} \ln\left(\frac{C_{OX}}{C_{RED}}\right) \qquad (3\text{-}4)$$

The *formal potential*, $E^{\circ\prime}$, replaces the first two terms on the right in Eq. (3-4), giving:

$$E = E^{\circ\prime} + \frac{RT}{nF} \ln\left(\frac{C_{OX}}{C_{RED}}\right) \qquad (3\text{-}5)$$

This is the appropriate form of the Nernst equation to use with concentrations rather than activities.

The formal potential is defined as the potential of the half-cell at unit concentrations of both OX and RED and at the specified level of supporting electrolyte. This is an operational definition and includes whatever factors influence the potential. The term $E^{\circ\prime}$ changes only slightly with the concentrations of OX and RED, since the activity coefficients depend primarily on the supporting electrolyte. A selection of formal potentials is given in Table 3-3 and in Figure 3-4. It is characteristic of formal potentials that the effect of species other than OX and RED, including complexing agents, are already taken into account for the given solution and need not be included explicitly in the Nernst equation.

TABLE 3-3
Formal Potentials at 25°C [13]
(vs. SHE)

Half Reaction	Supporting Electrolyte	Potential
$Ag^+ + e^- = Ag$	$HClO_4$, 1 M	+0.79
$Ag^+ + e^- = Ag$	H_2SO_4, 1 M	+0.77
$H_3AsO_4 + 2H^+ + 2e^- = HAsO_2 + 2H_2O$	$HClO_4$, 1 M	+0.58
$Ce(IV) + e^- = Ce(III)$	$HClO_4$, 1 M	+1.70
Same	H_2SO_4, 1 M	+1.44
$Co(III) + e^- = Co(II)$	H_2SO_4, 8 M	+1.82
$Cr(III) + e^- = Cr(II)$	$CaCl_2$, sat'd	−0.26
Same	H_2SO_4, 0.5 M	−0.37
$CrO_4^{--} + 2H_2O + 3e^- = CrO_2^- + 4OH^-$	NaOH, 1 M	−0.12
$Cr_2O_7^{--} + 14H^+ + 6e^- = 2Cr^{+++} + 7H_2O$	H_2SO_4, 0.1 M	+0.92
Same	$HClO_4$, 1 M	+1.03
$Cu(II) + e^- = Cu(I)$	NH_3, 1 M + NH_4^+, 1 M	+0.01
$Fe(III) + e^- = Fe(II)$	$HClO_4$, 1 M	+0.74
Same	H_2SO_4, 1 M	+0.68
Same	HCl, 0.5 M	+0.71
$Fe(CN)_6^{---} + e^- = Fe(CN)_6^{----}$	HCl, 1 M	+0.71
$Mn(III) + e^- = Mn(II)$	H_2SO_4, 7.5 M	+1.50
$Ni(CN)_4^{--} + e^- = Ni(CN)_4^{---}$	KCN, 1 M	−0.82
$Pb(II) + 2e^- = Pb$	NaOAc, 1 M	−0.32
$Sn(IV) + 2e^- = Sn(II)$	HCl, 1 M	+0.14
$Tl(III) + 2e^- = Tl(I)$	HCl, 1 M	+0.78

CLASSIFICATION OF ELECTRODES

Electrodes can be categorized according to the basic chemistry responsible for the potential. A metal in equilibrium with a solution of its ions forms a *Class I electrode*. An example is a copper foil in contact with a solution of cupric ions. The potential is given by the Nernst equation if no interfering species are present. Unfortunately, Class I electrodes are frequently affected also by the redox potential of the solution, which impairs their usefulness.

If the electrode material is electrochemically inert, it will not have a definite Class I type potential, but will continue to respond to the redox potential of the solution, forming a *redox electrode*. An example is a platinum wire in a solution containing both Fe^{+++} and Fe^{++} ions.

A second type of electrode consists of a metal in equilibrium with a sparingly soluble salt of the same element. This is a *Class II electrode*, examples of which are

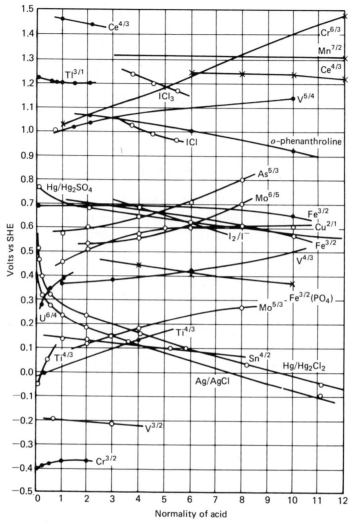

Figure 3-4. Formal reduction potentials of various systems, relative to the SHE. The notation $M^{4/3}$ denotes the potentials pertaining to the equilibrium $M(IV) + e^- = M(III)$. Open circles denote chloride solutions, full circles sulfates, and crosses phosphates. (After Furman [29] with the addition of curves for $Cr^{6/3}$, $Mn^{7/2}$ and $Ce^{4/3}$ phosphates, after Rao and Rao [30].)

the calomel and Ag/AgCl reference electrodes. They are somewhat selective in their response to the concentration of the *anion*. This can be understood if we replace the activity of the metal ion in the Nernst equation by its value obtained from the solubility product. For the case of the silver chloride electrode, $a_{Ag^+} = K_{sp}/a_{Cl^-}$ and:

$$E = E° + 0.0592 \log a_{Ag^+} = E° + 0.0592 \log K_{sp} - 0.0592 \log a_{Cl^-} \quad (3-6)$$

Thus, the electrode responds selectively to the presence of Cl⁻ ion, but any other species that affects the activity of Ag^+ ion, such as Br^- or NH_3, will interfere.

Another type is the *membrane electrode*, where the potential is developed across a membrane separating an internal reference from the solution of interest. The glass electrode and other ion-selective electrodes fall in this category. Finally, a rather elusive group includes ion-selective field-effect transistors (ISFETs) used as chemical sensors. These are usually considered to be a subclass of ion-selective electrodes.

THE GLASS ELECTRODE

The most widely used indicator is the glass electrode, encountered in almost every laboratory. It represents the method of choice for measuring pH.

The glass electrode consists of a thin glass bulb of carefully controlled composition, provided with an inside reference electrode in a solution of fixed pH and chloride ion content. The ensemble is shown in Figure 3-5. In use, the electrode is dipped in the solution of interest together with a second, external, reference electrode. The potential follows the equation:

$$E = E_{ref(int)} - E_{ref(ext)} + E_{asym} + 0.0592 \, (pH_{int} - pH_{ext}) \qquad (3\text{-}7)$$

In this equation, all terms are constant except pH_{ext} and we can write:

$$E = (Const) - 0.0592 \, pH \qquad (3\text{-}8)$$

The term E_{asym} describes the experimentally observable potential between the two sides of the membrane when the pH is the same on both sides. This potential is

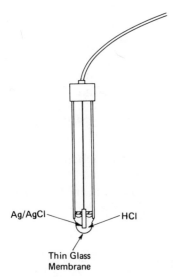

Ag/AgCl

HCl

Thin Glass
Membrane

Figure 3-5.   The construction of a glass electrode.

assumed to remain unchanged even with solutions of different pH. This *asymmetry potential* is caused by differences in the structure and composition of the inner and outer surfaces of the glass membrane. It originates as some constituents volatilize from the outside of the bulb during fabrication. It continues to change slowly in use, as the electrode becomes more or less hydrated, dehydrated, etched, and contaminated by various solutions. The effect may amount to perhaps as much as one pH unit, and this precludes the use of the glass electrode for absolute potentiometry.

In measurement, standard buffers are used to validate the pH scale of the meter. For moderate precision, a single buffer method can be used, according to the relation:

$$pH_{unkn} = pH_{buff} + \frac{E_{buff} - E_{unkn}}{0.0592} \qquad (3\text{-}9)$$

The factor 0.0592 is the *slope*, S, of the calibration plot at 25°C. If higher accuracy is desired, the slope cannot be trusted to have precisely this ideal value, and a second buffer must be used to determine the correct value, of S:

$$S = \frac{E_{buff\,2} - E_{buff\,1}}{pH_{buff\,1} - pH_{buff\,2}} \qquad (3\text{-}10)$$

Consequently, the expression to use is:

$$pH_{unkn} = pH_{buff} + \frac{E_{buff} - E_{unkn}}{S} \qquad (3\text{-}11)$$

In practice, the pH meter provides only the adjustment required for use with a single buffer and assumes the slope to be 0.0592 V per pH unit, called the *nernstian slope*. For more precise work, or if the electrode is suspected of being substandard, the control usually marked "temperature compensator" can be used to alter the slope just enough to give consistent readings for two buffers without regard for the actual temperature [14]. This should permit correct readings across the scale if the temperature of the buffers and sample are equal.

The procedure outlined above is consistent with the operational definition of pH as given by IUPAC [15]. Values of pH have been assigned to carefully selected buffers, and Eqs. (3-10) and (3-11) are used to obtain the correct pH for any solution. A hydrogen electrode is specified by IUPAC, but a glass electrode gives almost identical results. The reference buffers specified for the U.S. by the National Bureau of Standards are given in Table 3-4, together with the British standard and some secondary buffers. For most analytical purposes, somewhat less precise buffers are satisfactory, in view of the fact that the standard deviation inherent in a combination glass and reference electrode [16] is about 0.02 pH unit.

The behavior of the glass electrode depends on the ratio of various major constituents of the glass, as well as on the presence of minor species. Its active com-

TABLE 3-4
U. S. National Bureau of Standards pH Reference Buffers [15][a].

Composition (molality, m)	pH at $25°C$
Primary References	
KH Tartrate, saturated at $25°C$	3.557
KH_2 Citrate, 0.05	3.776
KH Phthalate, 0.05	4.004
$KH_2PO_4 + Na_2HPO_4$, each 0.025	6.863
KH_2PO_4, 0.008695 + Na_2HPO_4, 0.03043	7.415
$Na_2B_4O_7$, 0.01	9.183
$NaHCO_3 + Na_2CO_3$, each 0.025	10.014
Secondary References	
KH_3 Oxalate, 0.05	1.679
$Tris^b$, 0.1667 + Tris-HCl, 0.05	7.699
$Ca(OH)_2$, saturated at $25°C$	12.454

[a] Notes: The standard in the United Kingdom uses only one solution, 0.05
M (not m) KH-phthalate, to which a pH of exactly 4 is assigned.
[b] Tris = tris(hydroxymethyl)aminomethane.

ponent is an alkali ion, usually Na^+, which is responsible for the transport of electricity across the membrane. The activity of sodium ion is defined at each surface by an ion-exchange reaction of the type:

$$H_{soln}^+ + Na_{surf}^+ = H_{surf}^+ + Na_{soln}^+ \tag{3-12}$$

The situation has some similarity to the behavior of Class II electrodes, where the activity of the primary ion is determined by the activity of a second ion. This relation is shown in Figure 3-6. If equilibrium is assumed, the overall process can be thought of as the transport of Na^+ ions from a region of activity proportional to $a_{H^+(int)}$ to an area where the activity is governed by $a_{H^+(ext)}$. Since the transport is entirely carried by positive ions, an electrical potential will appear as the positive charges accumulate at the outer wall, the negatively charged silicate ions remaining stationary. The equilibrium is then electrochemical (elchem), rather than strictly chemical, so that the energy relations at equilibrium will be given by:†

$$\Delta G_{elchem} = \Delta G_{chem} + nFE \tag{3-13}$$

†In the presence of electrical potentials, the free energy of a reaction contains an electrical component equal to nFE joules per mole.

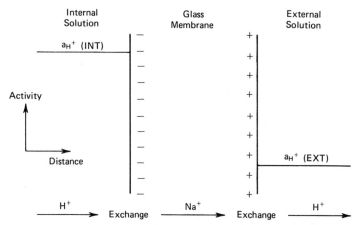

Figure 3-6. Simple model for the glass electrode. It is assumed arbitrarily that the activity of the hydrogen ion in the internal solution is greater than that in the outer solution. The reference electrodes on the two sides of the membrane are taken as identical.

Equilibrium occurs when $E = -\Delta G_{\text{chem}}/nF$, or, making use of the relation between free energy and activity, we can write:

$$E = -\frac{RT}{nF} \ln \frac{a_{\text{Na}^+(\text{ext})}}{a_{\text{Na}^+(\text{int})}} = -\frac{RT}{nF} \ln \frac{a_{\text{H}^+(\text{ext})}}{a_{\text{H}^+(\text{int})}} \tag{3-14}$$

If the ratio of sodium ion activities is not exactly equal to the ratio of hydrogen ion activities, a contribution to the asymmetry potential will result.

Since the glass electrode depends for its functioning on the exchange of alkali ions with protons, the exchange is less than complete at very low activity of hydrogen ion and at high alkali ion activity, the potential depends to some extent on the activity of Na^+ and other alkali ions present. In the case of sodium, the effect is taken into account by the *Eisenman equation:*

$$E = (\text{const}) + 0.0592 \log (a_{\text{H}^+} + k_{\text{pot}} a_{\text{Na}^+}) \tag{3-15}$$

where k_{pot} is the *sodium selectivity coefficient*, a very small quantity, typically less than 10^{-12}. Its effect is felt only beyond pH 10 or 11. A good pH electrode exhibits a sodium error of only about -0.2 pH unit for $a_{\text{Na}^+} < 1\ M$. (The minus sign implies that the reading is too low.)

Another type of deviation appears at the acid end of the scale. In reality, two different processes are involved here. The first is a deterioration of the surface of the glass in the presence of the acid; HCl is particularly damaging. The effect is progressive, and one can no longer obtain stable readings. The extent of the error could be as much as 0.5 pH unit or even more. A second type of deviation is due to the severe alteration of the activity coefficient of the hydrogen ion in solutions of high ionic strength. For example, in 15 M KBr the activity coefficient of the H^+ ion is about 4000. This might be suspected to be an artifact, but such large coefficients are observable both with glass and hydrogen electrodes, and it is highly improbable that the two are in error by the same amount, since they have such different mechanisms [17].

ION-SELECTIVE ELECTRODES

The glass electrode is only one of a large number of devices, called *ion-selective electrodes*, that exhibit a potential proportional to the logarithm of the activity of some specific ion. (The term "ion-specific" has also been used, but is not recommended [18], since the response is not restricted to only one ion, and in general, several ions contribute to the potential.)

Ion-selective electrodes have been known since the work of Kolthoff [19] in 1937, but have received little attention until the publications of Pungor in 1961. The development of the fluoride electrode in 1966 responded to the great demand for a convenient way to monitor the fluoride content of potable waters.

The construction of an ion-selective electrode is quite similar to that of the glass electrode, with a membrane enclosing an inner reference half-cell to be measured against an external reference. (The inner reference can sometimes be replaced by a metallic connection, as will be seen below.) The potential response, E, of the electrode is given by a modified Nernst equation [18] which takes into account that in addition to the principal species A of charge z_a, there are other ions, B, C, and so on, that contribute to the potential:

$$E = (\text{const}) + \frac{0.0592}{z_a} \log [a_a + k_{pot}^{A,B} (a_b)^{z_a/z_b} + k_{pot}^{A,C} (a_c)^{z_a/z_c} + \cdots] \quad (3\text{-}16)$$

The constants k_{pot} are the potentiometric selectivity coefficients and represent the relative effects of various ions on the potential. The choice as to which ion is the primary ion is somewhat arbitrary, and if A designates the ion for which the response is strongest, then the coefficients will be small numbers. Occasionally, the ion of interest might not be the dominant one, and then the selectivity coefficients would be large. A large k_{pot} means that the corresponding species can interfere with the measurement of the primary ion, and must be absent.

Selectivity coefficients have received much attention [20], but until recently there has been some confusion about the exact meaning of the concept. The coefficient is not a constant, but depends on the concentration of the interfering species. The preferred method for determing it is to make a series of measurements at various concentrations of the primary ion A with the interferent B at a constant level. The data can be presented in the form of a graph of the potential plotted against log a_a, as seen in Figure 3-7. The curve consists of two linear segments, one in which the electrode responds to the primary ion, characterized by a

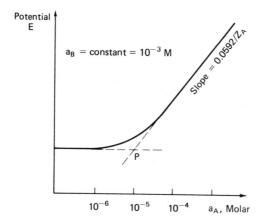

Figure 3-7. The measurements of $k_{pot}^{A,B}$. If $z_a = z_b$, the relation is: $k_{pot}^{A,B} = a_a/a_b$, which gives for point P the value $10^{-5}/10^{-3} = 0.01$.

slope of about $0.0592/z_a$, and a horizontal segment in which the electrode actually responds to the activity a_b of the interfering ion (which is held constant). The activities at point P can be used to calculate the selectivity coefficients by the relation:

$$a_a = k_{pot}^{A,B}(a_b)^{z_a/z_b} \tag{3-17}$$

In the figure, the charges z_a and z_b are taken to be equal, which is usually the case. Note, however, that if the two charges are not equal, one of the activities will be raised to a power, so that k_{pot} will have a widely different value for a given degree of interference. This can be misleading when the relative interferences between two ions of different charges are compared.

Even in the absence of interfering species, the sloping portion of the curve cannot continue indefinitely. This is because in the relation:

$$E = (\text{const}) + 0.0592 \log a_a \tag{3-18}$$

which applies in the absence of interferences, E tends to go toward minus infinity as the concentration of A approaches zero. This implies that *whatever is absent produces an infinite potential*, an interesting paradox. For our purposes, we need only keep in mind that Eqs. (3-16) and (3-18) are merely approximations, valid over restricted ranges. Outside this range, the electrode response takes the form indicated in Figure 3-8, which resembles Figure 3-7, but in the absence of B. The point P can be taken as the limit of detection.

Ion-selective electrodes, in common with the glass electrode, deviate more or less from the nernstian slope of $0.0592/z$, so that an experimental slope S must be determined. Asymmetry potentials on the order of perhaps 10 mV can also be present.

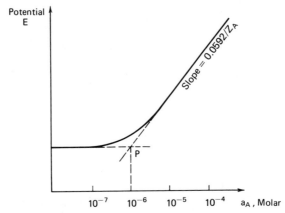

Figure 3-8.   Determination of the limit of detection, shown as 10^{-6} M.

Classification of Ion-Selective Electrodes

In terms of the mode of operation, ion-selective electrodes can be classified [18] as *primary*, those that respond to the species of interest directly, and *sensitized*, those in which the response is to some other species, not necessarily ionic, through the agency of a sensitizer such as an enzyme. Primary electrodes, in turn, can be classified as follows:

1. Crystalline membrane electrodes
   a. Homogeneous membranes
   b. Heterogeneous membranes
2. Noncrystalline membrane electrodes
   a. Rigid matrix
   b. Nonrigid matrix
      i.   Positively charged mobile carrier
      ii. Negatively charged mobile carrier
      iii. Neutral mobile carrier
3. Field-effect transistor sensors (ISFETs)

In the following paragraphs we will discuss these various types in more detail. For additional information, consult the references [21–25].

The active element of a *homogeneous crystalline membrane* electrode consists of a solid material like LaF_3, $AgCl$, Ag_2S, or a mixture such as $Ag_2Se + Cu_2Se$ or $Ag_2S + AgI$. The physical construction of a typical electrode assembly is shown in Figure 3-9a. The internal solution contains the primary ion at constant activity in contact with a reference electrode so that the electrode potential varies only with the activity of the ions in the external solution. It is interesting to note, though, that since the electrode generally responds to several ions (see Eq. 3-18), mean-

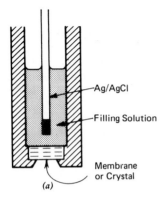

-Ag/AgCl

-Filling Solution

Membrane
or Crystal

(a)

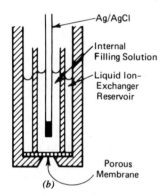

-Ag/AgCl

Internal
Filling Solution

Liquid Ion–
Exchanger
Reservoir

Porous
Membrane

(b)

Figure 3-9. Construction of two types of ion-selective electrodes. (a) Solid membrane; (b) liquid ion-exchanger.

ingful results can be obtained even if the internal and external solutions do not have any ions in common. In fact, it is possible under some circumstances to dispense completely with the internal solution and substitute a direct contact between a metallic conductor and the membrane. This implies that the metal/crystal interface is in thermodynamic equilibrium [26]. It appears that this metallic contact plays a more complex role than one would expect, since the nature of the metal influences the E° of the electrode. Thus, a Ag_2S membrane gives different potentials when the contact is with silver and with carbon [26].

A very useful example of the crystalline membrane type is the fluoride electrode, which utilizes a LaF_3 single crystal doped with europium or samarium. It exhibits a laudable selectivity, being seriously affected only by OH^-, so that solutions to be measured must be slightly acidified. This electrode is particularly important because the classical determination of fluoride is a very difficult matter.

The selectivity of crystalline membrane electrodes depends primarily on the solubility of the various salts involved. Thus, the AgCl electrode is sensitive to Br^- and I^-, since silver bromide and iodide are less soluble than the chloride. A few examples of homogeneous membrane electrodes are given in Table 3-5.

Heterogeneous membrane electrodes are prepared by incorporating the pow-

TABLE 3-5

Homogeneous Crystalline Membrane Selective-Ion Electrodes [20][a]

Primary Ion	Active Material	Major Interferences
F^-	LaF_3, single cryst.	OH^-
Cl^-, S^{--}	$AgCl$	$Br^-, I^-, CN^-, NH_3, S_2O_3^{--}, S^{--}$
Br^-	$AgBr$	$I^-, CN^-, NH_3, S_2O_3^{--}, S^{--}$
I^-	AgI	CN^-, S^{--}
CN^-	AgI	I^-, S^{--}
SCN^-	$AgSCN + Ag_2S$	S^{--}, I^-, Br^-
S^{--}	Ag_2S	Ag^+
Ag^+	Ag_2S	S^{--}
SO_3^{--}	$HgS + Hg_2Cl_2$	Cl^-, Br^-, I^-, SCN^-
Cd^{++}	CdS or $(CdS + Ag_2S)$	$H^+, Mn^{++}, Pb^{++}, Fe^{+++}, Cr_2O_7^{--}$
Pb^{++}	$PbS + Ag_2S$	Cu^{++}, Cd^{++}

[a]Note: Tables 3-5 through 3-12 are for illustration only; for specific details, consult the references.

dered active crystals into an inert matrix such as polyvinylchloride or silicone rubber. The flexible nature of the resulting membrane is advantageous, in that it resists breakage. Some examples of this class are given in Table 3-6. Of widespread use are silver halide membranes, which respond not only to the halide ions, but also to other ions that form sparingly soluble precipitates with Ag^+, such as S^{--}, or complexing agents like CN^-. Unfortunately, many selective-ion electrodes are even less specific. For example, the combination $CdS + Ag_2S$ membrane forms a cadmium

TABLE 3-6

Examples of Heterogeneous Crystalline Membrane Electrodes
[20, 23]

Primary Ion	Type	Major Inteferences
F^-	LaF_3 in silicone rubber	OH^-
Cl^-	$AgCl$ in silicone rubber	$S^{--}, I^-, Br^-, SCN^-, NH_3$
S^{--}	Ag_2S in silicone rubber	(None)
Cs^+	Cs tungstoarsenate in epoxy	(None)
Cd^{++}	CdS in polyethylene	Pb^{++}, Fe^{+++}
Tl^+	Tl molybdophosphate in epoxy	(None)
Cu^{++}	$Cu_2S + Ag_2S$ in thermoplastic polymer	Cu^+

electrode that is affected also by Ag^+, Fe^{+++}, Hg^{++}, and a number of other divalent cations.

The category of *rigid matrix noncrystalline electrodes* involves mainly silicate glass membranes. In addition to the conventional pH electrode, we include here alkali-sensitive units that can be optimized to be reasonably selective with respect to any one of the ions Li^+, Na^+, K^+ or NH_4^+. Reliable Ag^+-sensitive glasses have also been described. The physical construction is similar to that of the pH glass electrodes.

Positively charged mobile-carrier electrodes have received much attention, since they can be made sensitive to a variety of anions. Of some importance are the ClO_4^- and NO_3^- electrodes, even though they are rather poorly selective. In addition, electrodes have been made that respond to such other anions as trifluoro-acetate, maleate, and phthalate. A selection of such electrodes is listed in Table 3-7. These electrodes contain a liquid ion-exchanger, a substance that has the ability to enter into heterogeneous equilibrium with the ion of interest when dissolved in an organic solvent. A porous solid material is impregnated with the organic phase, thus forming a separating membrane between inner and outer solutions, as shown in Figure 3-9*b*. Extraction equilibria replace the solubility equilibria of the crystalline membranes in determining the selectivity.

A number of such electrodes have been constructed using derivatives of *o*-phenanthrolene and long-chain substituted ammonium ion compounds as active constituents. A principal disadvantage of this type of membrane is that it responds to a variety of species. In connection with some other procedure, such as titration, that ensures specificity, they can render excellent service.

Negatively charged mobile-carrier electrodes are similar to the previous category except that the ion-exchanger contains a bulky *anion* which makes the electrode sensitive to cations such as Ca^{++}, Mg^{++}, or K^+. The anion can be, for example, tetrachlorophenyl borate or didecyl phosphate (Table 3-8). Again, the selectivity is not very good, but on the other hand, mobile-carrier electrodes have relatively low resistance, which facilitates the miniaturization needed for biological purposes.

TABLE 3-7
Examples of Positively Charged Mobile-Carrier Electrodes [20]

Primary Ion	Type	Major Interferences
Cl^-, Br^-, I^-	Cetyl-$(CH_3)_3NOH$ in octanol	ClO_4^-, NO_3^-, OH^-, SO_4^{--}
I^-	Aliquat-336S on coated wire	NO_3^-
SCN^-	Aliquat-336S on coated wire	I^-, SO_4^{--}
ClO_4^-	Fe (o-phenanthrolene)$_3$ -$(ClO_4)_2$ in nitrobenzene	OH^-
NO_3^-	Cetyl-$(CH_3)_3NOH$	Cl^-, Br^-, SO_4^{--}, NO_2^-
Phenylalanine	Aliquat-336S in decanol	NO_3^-, Cl^-, other amino acids

Notes: (1) Aliquat-336S = methyltricaprylammonium salt. (2) The list of interferences is not exhaustive.

TABLE 3-8
Examples of Negatively Charged Mobile-Carrier Electrodes [20]

Primary Ion	Membrane	Major Interferences[a]
Ca^{++}	Didecylphosphoric acid in Di-n-octylphenylphosphonate	H^+, Ba^{++}, Mg^{++}
Zn^{++}	Zn salt of di-n-octyl-phenylphosphoric acid	Mg^{++}, Ca^{++}, Sr^{++}, Cd^{++}, Pb^{++}, Ba^{++}
K^+	Ion exchanger (Corning)	Cs^+, Rb^+, Na^+, NH_4^+, Ca^{++}

[a]Note: There are numerous other interferences.

Both positive and negative mobile-carrier electrodes have been adapted to metallic contact by simply coating a wire with the active material. Such *coated-wire electrodes* are found to operate satisfactorily, while providing the advantages of low cost, simplicity, and ruggedness [27]. They suffer, however, from limited lifetimes.

Neutral mobile-carrier electrodes are based on a large neutral molecule that can complex the ion of interest to form an aggregate that is soluble in an organic solvent. The solution so formed is then loaded into a porous solid to constitute the ion-selective membrane. The carrier can be a macrocyclic compound as shown in Figure 3-10a, or an open chain compound, as in (b). In both cases, ion-dipole interactions between the oxygen atoms and the ion are responsible for the stability of the complex. Many neutral carrier molecules have biological activity, since they can effect the transport of alkali ions across biological membranes. Some examples are given in Table 3-9.

Sensitized electrodes constitute a very important group because of their inherent selectivity, as *two* membranes are used to generate the response, each with its own

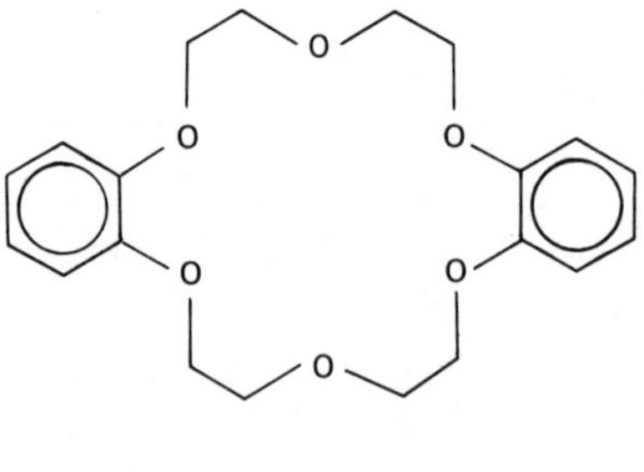

(a) 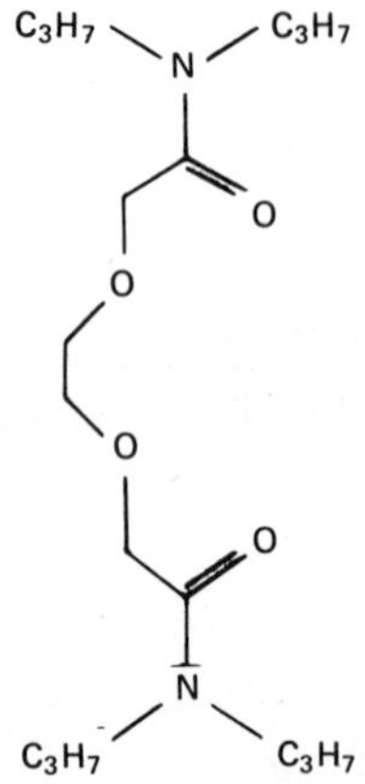   (b)

Figure 3-10.   Examples of neutral carriers. (a) Macrocyclic (crown-6 type); (b) noncyclic.

TABLE 3-9

Examples of Neutral Carrier Ion-Selective Electrodes [20]

Ion	Carrier	Interferences
K^+	30-Crown-10 derivative in PVC	Rb^+, Cs^+, Ca^{++}
K^+	Valinomycin (an antibiotic) in diphenyl ether	NH_4^+, Rb^+, Cs^+
Ca^{++}	Antibiotic A-23187	Sr^{++}, Na^+, Mg^{++}
Ba^{++}	Polyethyleneglycol derivative	Sr^{++}

selectivity characteristics. One of the oldest of such devices is a CO_2-sensitive probe which consists of a glass pH electrode in contact with a thin layer of $NaHCO_3$ solution (Figure 3-11). The outer membrane, made of an organic polymer, is hydrophobic so that aqueous liquids cannot penetrate it. The bicarbonate solution is located in the thin space between the polymer and glass membranes. The ensemble can be used to measure the CO_2 content of an external gas, or it can be immersed in a solution to be analyzed; in the latter instance, an airgap is retained in the pores of the membrane, allowing the passage of the CO_2. The electrode has a few interferences, such as SO_2, NH_3, acetic acid, and, interestingly, water. Water interferes if the osmotic pressures on the two sides of the membrane are different, since water vapor will then cross the membrane, changing the concentration of the bicarbonate layer. This difficulty can be eliminated by the use of an osmotic

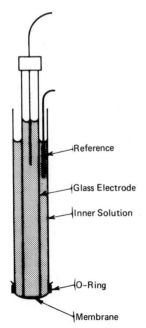

Reference

Glass Electrode

Inner Solution

O-Ring

Membrane

Figure 3-11. Gas-sensing probe using a glass electrode.

TABLE 3-10

Examples of Gas-Sensing Electrodes [28]

Species Sensed	Lower Limit (M)	Sensor Ion	Inner Solution
CO_2	10^{-5}	H^+	$0.01\ M$ $NaHCO_3$
NH_3	10^{-6}	H^+	$0.1\ M$ NH_4Cl
H_2S	10^{-8}	S^{--}	Buffer, pH = 5
Cl_2	10^{-3}	Cl^-	Buffer, HSO_4^-

strength adjuster to equalize the pressures. The CO_2 electrode exhibits a nernstian response with respect to $\log p$. Similar electrodes have been made for various other gaseous species, as shown in Table 3-10.

Another possibility is to use an enzyme-containing membrane between the solution and the glass electrode (Table 3-11). This will react with the species of interest to generate a substance that can be sensed by the electrode. For example, a membrane containing urease will decompose urea to form NH_3 to which the inner electrode is sensitized. Such *enzyme electrodes* have the outstanding selectivity characteristic of enzyme systems. They tend to suffer from short lifetimes (a few weeks) and slow speed of response.

Ion-sensitive field-effect transistors (*ISFETs*) are relatively new on the scene, first reported in 1970. They have been found useful [25-29], especially for miniaturized sensors. ISFETs can be described as hybrid electronic devices consisting of an ion-selective membrane and a field-effect transistor (FET) preamplifier in a single unit. An ISFET can be visualized as constructed by the first stage of an electronic voltmeter placed into a capsule in contact with the membrane to form a preamplifier probe (Figure 3-12). The fundamental advantage lies in the improvement in the signal-to-noise ratio, since much of the noise in an electronic instrument is generated by the wiring of the input stage.

The structure of a typical insulated-gate FET is shown in Figure 3-13a, in a simplified form. The current flowing between the source and drain, through the

TABLE 3-11

Examples of Enzyme Electrodes [21, 22]

Species Sensed	Enzyme	Sensor
l-Arginine	Arginine carboxylase	CO_3^-
Glucose	Glucose oxidase (generates H_2O_2)	I^-
Amygdaline	β-glucosidase	CN^-
Urea	Urease	NH_3
NO_2^-	Nitrite reductase	NH_3

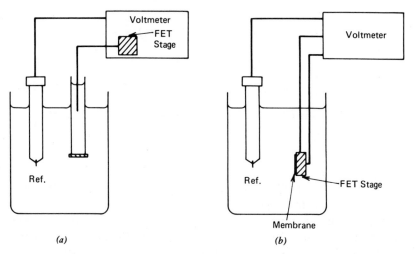

(a)   (b)

Figure 3-12.   Relation between a FET stage in voltmeter or pH-meter and an ISFET [25].

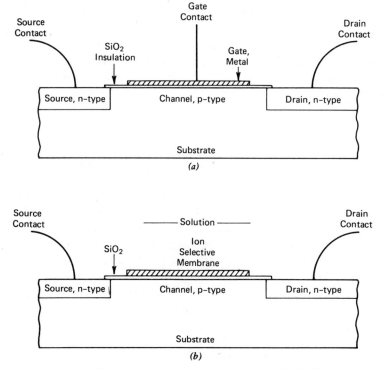

Figure 3-13.   (a) A simplified representation of an insulated-gate field-effect transistor (FET). (b) An ISFET derived from it. The charge carriers are positive for a p-type channel.

TABLE 3-12
Examples of ISFETs [25]

Ion	Membrane
H^+	Hydrated silica
Cl^-	AgCl
I^- and CN^-	AgI + AgCN
K^+	Valinomycin in solvent

channel, is carried by positive charges ("holes"). The potential applied to the gate cannot produce a current itself because of the insulating layer of SiO_2, but it affects the distribution of carriers and thus the source-to-drain resistance. In other words, the current is a function of the potential applied to the gate. The modification needed to form an ISFET is replacing the gate of the FET by an ion-selective membrane (the hydrated surface of the SiO_2 insulating layer itself may serve as the ion-selective element). The drain current of the FET, as controlled by the ion-selective membrane, is then sent by wires to the meter. A few examples of such devices are presented in Table 3-12.

INSTRUMENTATION

The measurement of the potential of an electrode is normally carried out under conditions approximating thermodynamic equilibrium, which implies that only a very small current is allowed to be drawn from the cell. To the extent to which current is actually passing, an error ΔE will appear, depending on the faradaic and ohmic resistance of the electrodes:

$$\Delta E = IR_{far} + IR_{ohmic} \qquad (3\text{-}19)$$

(The double-layer impedance need not be taken into account unless very fast measurements are performed.)

In general, the contribution of the reference electrode to the resistance of the cell is less than 1000 ohms, and the dominating factor is the resistance of the indicator. For ion-selective electrodes, the total resistance can be as high as 10 megohms or so; in a few cases, notably the glass electrodes, it is occasionally much higher. For a resistance of 10^7 Ω, Eq. (3-19) indicates that the current must be less than 10^{-10} A if ΔE is to be kept below 1 mV. Consequently, a conventional voltage follower might give substantial errors. Electrometer voltage followers designed for this service are generally called pH meters or selective-ion meters.

REFERENCES

1. See Refs. 2, 3, and 4 of Chapter 2.
2. R. Gomer and G. Tryson, *J. Chem. Phys.*, 1977, *66*, 4413.

3. V. D. Parker, *J. Am. Chem. Soc.*, 1974, *96*, 5656.
4. I. Zagorska and Z. Koczorowski, *J. Electroanal. Chem.*, 1979, *101*, 317.
5. H. S. Harned and R. W. Ehlers, *J. Am. Chem. Soc.*, 1933, *55*, 2179.
6. K. K. Kundu, D. Jana and M. N. Das, *J. Phys. Chem.*, 1970, *74*, 2625.
7. D. J. G. Ives and G. J. Janz, "Reference Electrodes," Academic Press, New York, 1961.
8. J. H. Hildebrand, *J. Am. Chem. Soc.*, 1913, *35*, 847.
9. R. G. Bates, "Determination of pH: Theory and Practice," Wiley, New York, 1964.
10. E. A. Guggenheim, *J. Phys. Chem.*, 1930, *34*, 1758.
11. P. A. Rock, *J. Chem. Educ.*, 1970, *47*, 683.
12. H. J. Hickman, *Chem. Eng. Sci.,* 1970, *25*, 381.
13. L. Meites, "Handbook of Analytical Chemistry," McGraw-Hill, New York, 1963.
14. W. J. Eilbeck, *J. Chem. Educ.*, 1980, *57*, 834.
15. R. A. Durst and J. P. Cali, *Pure Appl. Chem.*, 1978, *50*, 1485.
16. S. Ebel, E. Glaser and H. Mohr, *Fresenius' Z. anal. Chem.*, 1978, *293*, 33.
17. K. Schwabe, pH Measurements and Their Applications, in "Electroanalytical Chemistry" (H. W. Nürnberg, ed.), Wiley-Interscience, New York, 1974, p. 536.
18. G. G. Guilbault, *Pure Appl. Chem.*, 1976, *48*, 127.
19. I. M. Kolthoff and H. L. Sanders, *J. Am. Chem. Soc.*, 1937, *59*, 416.
20. E. Pungor, K. Toth and A. Hrabeczy-Pall, *Pure Appl. Chem.*, 1979, *51*, 1913.
21. J. Koryta, *Anal. Chim. Acta*, 1972, *61*, 329.
22. J. Koryta, *Anal. Chim. Acta*, 1977, *91*, 1.
23. R. P. Buck, *Anal. Chem.*, 1978, *50*, 17R.
24. G. H. Fricke, *Anal. Chem.*, 1980, *52*, 259R.
25. J. Janata and R. J. Huber, *Ion Selective Electrode Rev.*, 1979, *1*, 31.
26. R. P. Buck and V. R. Shepard, Jr., *Anal. Chem.*, 1974, *46*, 2097.
27. C. R. Martin and H. Freiser, *J. Chem. Educ.*, 1980, *57*, 512.
28. J. W. Ross, J. H. Reisman and J. A. Krueger, *Pure Appl. Chem.*, 1973, *36*, 473.
29. N. H. Furman, Potentiometry, in "Treatise on Analytical Chemistry" (I. M. Kolthoff and P. J. Elving, eds.), pt. I, vol. 4; Wiley-Interscience, New York, 1963, p. 2294.
30. G. G. Rao and P. K. Rao, *Talanta*, 1963, *10*, 1251; 1964, *11*, 825.

Chapter 4

VOLTAMMETRY:
I. POLAROGRAPHY

INTRODUCTION

Chapters 4 through 8 treat of voltammetry, the study of current-voltage relationships at a working electrode. The most extensively used voltammetric method is polarography, in which the working electrode takes the form of a series of tiny drops of mercury issuing from a fine glass capillary, the *dropping mercury electrode*. In this chapter, some basic features of voltammetry are discussed, and the general principles of polarography are developed with emphasis on the classical (DC) approach to this subject. It will serve as a foundation on which to build, when, in subsequent chapters, more advanced forms of polarography and voltammetry are treated.

VOLTAMMETRY

In voltammetry an electron-transfer reaction can take place at the electrode if the potential is appropriate, its extent being determined by the surface concentration of some electroactive species. The resulting current will be only a transient, decaying rapidly to zero, unless some mechanism is present to bring a continuously renewed supply of the electroactive material to the surface. A similar mechanism must generally be available to remove the product of reaction from the surface.

Hence, the subject of voltammetry must treat, not only the electron-transfer process itself, but also the detailed mechanisms of mass transport. This requires knowledge of the geometry of the cell and of the working electrode. In this section we explore the interdependence of these factors, and subsequently show how they can be applied in a practical sense in polarographic and other analytical techniques.

Two major modes of transport can be distinguished, due to: (1) convective motion of the solvent (and supporting electrolyte), carrying the active species with it, and (2) movement of the active species through the solvent, as by diffusion. Both of these modes are acting all the time that a current is passing through a cell, but usually one mode predominates.

In practical terms, two kinds of voltammetric experiments are commonly encountered, conducted with or without mechanical stirring. If the solution is stirred, clearly the resulting bulk motion of the liquid is the dominant transport mechanism, but even here diffusion plays a part. It is customary to hypothesize a thin stationary layer of solution in contact with the electrode, that is unaffected by stirring; the mechanism of transport across this thin layer is by diffusion.

In the other limiting case, the cell is protected as far as possible from mechanical forces that might cause convective motion of the solution. Here, the transport of active species is by *diffusion*, in which the motion is caused by a gradient in chemical potential, or by *migration*, the effect of an electrical potential gradient. Even in this circumstance, however, mechanical convection is never entirely absent, the residual effect being due to such sources as temperature changes and vibration. In some situations, the faradaic effects of the current may produce a solute of different density, and this causes some convection. The growth and fall of the mercury drops in polarography also creates a small convective effect.

Of all these mass transport mechanisms, diffusion and migration are most readily susceptible to detailed theoretical treatment. In the presence of supporting electrolyte, migration becomes of negligible importance.

It is instructive to consider the nature of the current response to a step potential function in the two limiting cases previously discussed. The corresponding current-time curves are plotted in Figure 4-1. The working electrode is specified only in that it is chemically inert and of fixed area. Zero time is taken as the instant when the potential is changed from well below to well above the formal potential $E^{\circ\prime}$ for an electroactive species. In an unstirred solution, the surface concentrations change rapidly in the sense necessary to diminish the current. This is given by the Cottrell

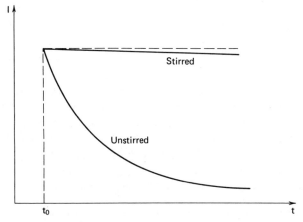

Figure 4-1. Curves of current against time for sitrred and unstirred solutions. The slope of the upper curve is exaggerated for clarity.

equation previously described [Eq. (3-26)]:

$$I = nFA(D/\pi t)^{1/2} C^*$$ (4-1)

Thus the current drops off as the inverse square root of the time. (The symbol C^* represents the bulk concentration of the active species.)

In the case of a well-stirred solution, fresh active material is always available to the electrode, so the current stays at a nearly constant value. The magnitude of the current is proportional to concentration, which diminishes slowly because of the charge-transfer reaction, and the curve drops exponentially with time at a slow rate.

In the absence of stirring, there is no noticeable change of bulk concentration, simply because during all but the first small fraction of the time, the current is very small (Figure 4-1). By contrast, if the solution is stirred, the current continues at a higher level, and appreciable changes in concentration may ensue.

One of the most useful electrodes for the study of voltammetry in unstirred solutions is the *dropping mercury electrode (DME)*, which consists of a continuing series of minute droplets of mercury issuing under pressure from a glass capillary. Voltammetry at the DME, called *polarography*,† is the principal subject of the present chapter.

Voltammetric methods can, in principle, be approached in either of two ways: the *current* can be recorded as the voltage is controlled, or alternatively, the *voltage* is monitored as a current is impressed. This complementary relation suggests that the phenomena involved are, in essence, concerned with the *impedance* between the electrode and the solution.

POLAROGRAPHY

Polarography was invented by Jaroslav Heyrovský and his students at the Charles University in Prague, with their initial publications in 1925 [1, 2]. Since then it has been developed extensively, particularly with the application of modern electronic technology.

The dropping mercury electrode, with its requirement of an elevated reservoir of mercury and other appurtenances, is rather a nuisance, but its inconvenience is more than outweighed by several major advantages. Because of the frequent dropwise renewal of the electrode, the concentration of any reduced metals dissolved in the mercury can never reach significant levels; also contamination of the electrode surface by adsorption cannot accumulate beyond the life of each drop. In common with all forms of mercury electrodes, the DME exhibits high overvoltage with respect to the reduction of water to hydrogen. This latter feature makes accessible to polarography ions of many metals with reduction potentials

---

†It must be noted that some authors and some manufacturers choose to use the term "polarography" in the broader sense, practically synonymous with "voltammetry," referring, for example, to "polarography at a platinum electrode."

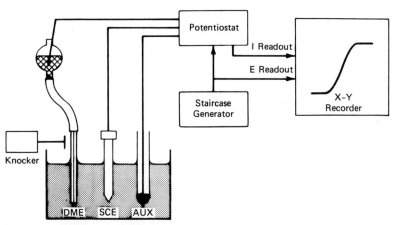

Figure 4-2. A simple polarograph with a three-electrode cell containing a Pt-wire anode (AUX), a DME cathode, and an SCE as reference.

in the range extending up to about −2.0 V versus SCE, that could never be observed at a platinum electrode. (However, platinum is advantageous at more positive potentials where mercury would be oxidized.)

The Basic Experiment

Consider an unstirred, three-electrode cell (Figure 4-2) connected to a potentiostat. The working electrode is the DME; the reference is assumed for this discussion to be a saturated calomel electrode (SCE). The auxiliary (counter) electrode can be a mercury pool or a platinum wire. The DME capillary is connected by a flexible tube to an elevated reservoir of mercury. The electronic circuitry is so designed that the potential advances stepwise, each drop being operated at a potential some 10 mV more negative than the previous drop. Each drop is knocked off mechanically at the same moment that the voltage is incremented, so that the drop is maintained at a constant potential throughout its lifetime.

Let us suppose that the cell contains a 0.5 M solution of KNO_3 saturated with air. The current passing through the cell is recorded as a function of the voltage. The resulting graph is shown in Figure 4-3a, curve 1. The deep serrations are the result of successive drop falls, in which the surface area, and hence the current, goes almost to zero. By an added electronic component called a *sampling circuit*,† the serrations can be eliminated, giving curve 2, in which a series of steps correspond to sequential drops. This is desirable for simplicity, since no additional information is carried by the lower portions of the serrations.

The waves in these curves are due to the cathodic reduction of dissolved oxygen, as can be shown by running similar curves after flushing out the oxygen with an inert gas, such as purified N_2 or Freon-12 [3], a process known as *sparging*. Figure 4-3b demonstrates the change in the curve resulting from oxygen removal. (Be-

†The sampling circuit is sometimes referred to as a *Tast* circuit.

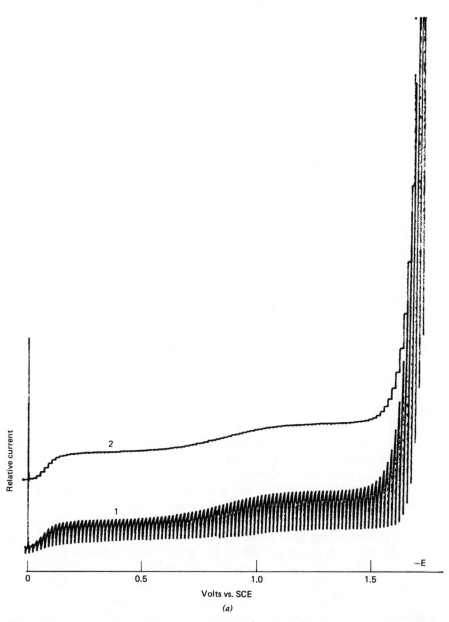

Figure 4-3. (*a*) Polarogram of supporting electrolyte (0.5 *M* KNO$_3$) in the presence of air: curve 1, conventional recording, curve 2, sampled (moved upward for clarity). (*b*) The effect of sparging with Freon-12 for various time intervals as marked; solution: 0.5 *M* KCl. (Recorded on a Sargent-Welch Polarograph.)

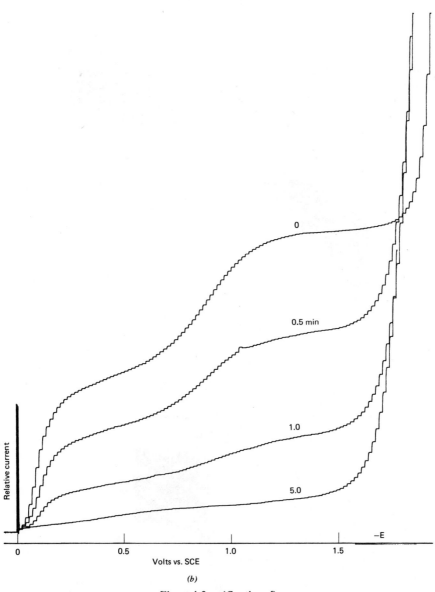

Figure 4-3. (*Continued*)

cause of this effect, it is standard practice to sparge all solutions routinely before running polarograms.)

If now the amplifier gain be increased, a response such as curve 1 (unsampled) of Figure 4-4*a* will result. Since no electroactive substance is present, the curiously shaped serrations must be due solely to the double-layer charging effects. The marked change at about −0.57 V coincides with the PZC, the point of zero charge,

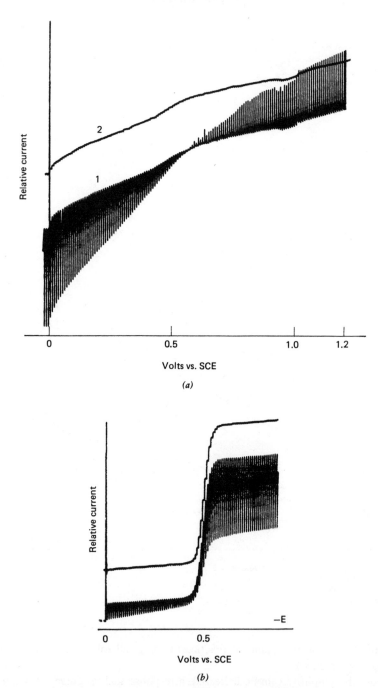

Figure 4-4. Typical DC polarograms. (*a*) Curve 1: unsampled background current; Curve 2, same, sampled. (*b*) 10^{-4} *M* Cd^{++} in 0.5 *M* KCl, sampled and unsampled. Sparged with Freon-12. Drop-time mechanically regulated at 1 s. Scan rate, 0.5 V/min. (Recorded on a Sargent-Welch Polarograph.)

where there is no tendency to form a charged double layer, and hence where no current passes. At points on either side of this, transient currents flow to charge the capacitance of the double-layer, but with opposite signs. As each drop falls from the capillary, the accumulated charge is abruptly lost and a new transient commences. If this is sampled just prior to each drop fall, curve 2 is seen, which is effectively the *upper* envelope of the more positive segment and the *lower* envelope of the more negative portion of the curve.

A complicating factor is that, as the voltage across the interface is scanned, the double-layer capacitance per unit area changes, being approximately 20 $\mu F/cm^2$ at voltages more negative than the PZC, and about twice this value at more positive potentials (p. 98 of Reference 4). This causes a corresponding variation in the charging current.

If an electroactive substance, such as Pb^{++} is present, the curves become modified as shown in Figure 4-4b. An abrupt step (usually called a "wave") appears at a voltage near to the formal reduction potential. In the region more negative than about -0.4 V in this example, a new component of current is evident, originating in the faradaic reduction of lead ions to the free metal. It may be noted that the sampled curve now corresponds to the *upper* envelope of the unsampled counterpart over the whole range.

An important innovation in polarographic instrumentation was announced in 1979 by Princeton Applied Research Corporation [5]. This firm introduced a mercury drop electrode assembly in which the mercury flow can be interrupted by a fast-acting valve under the control of an electronic timing circuit. The mercury is allowed to flow for a short period, such as 100 ms, following which the drop remains at constant size, hanging from the tip of the capillary by surface tension until dislodged and replaced. The advantage of this system (called the SMDE, for Stationary Mercury Drop Electrode), is that a measurement can be made on a drop of constant surface area. This avoids the difficulties inherent in the large periodic double-layer charging current of the conventional electrode, while preserving the advantage of renewing the mercury drop. Figure 4-5 shows the graph of drop area against time for the two types of electrodes.

The mathematical theory, as well as practical operating procedures, are con-

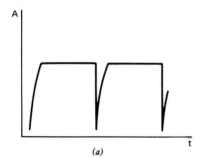
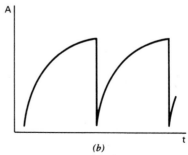

Figure 4-5.   The area of mercury-drop electrodes as a function of time. (*a*) Stationary mercury drop (Princeton Applied Research Corporation). (*b*) Conventional DME.

siderably simplified by substitution of the SMDE for the traditional DME. If the use of this device and comparable ones that may be introduced by other manufacturers becomes widespread, then the discussion of polarography in a book such as this will be rather different [6]. However, such an approach would be premature at the present time. Hence, the methods pertinent to the DME are here given primary emphasis with comments about the modifications required for the SMDE, where appropriate.

Now let us turn to the chemistry taking place at the electrode. The course of the faradaic current can be explained most conveniently if we consider the generalized reduction $OX + ne^- = RED$, in which the reduced species is soluble in mercury. The potential is given by the equation:

$$E = E^{\circ\prime} + \frac{0.0592}{n} \log \left(\frac{C_{OX}^s}{C_{RED}^s} \right) \qquad (4\text{-}2)$$

where the concentrations (superscript s) are taken as surface values on either side of the interface, the reduced metal being on the mercury side.

Let us consider for purposes of illustration a system such as Zn^{++}/Zn, for which $E^{\circ\prime} = -1.00$ V (vs. SCE), and for which $n = 2$. We can calculate the ratio of concentrations C_{OX}^s/C_{RED}^s that must exist at the electrode surface for any applied potential E [Eq. (4-2)]; these are presented in Table 4-1 for a series of potentials. We can see clearly from these figures that when the potential is further removed than about 100 mV from $E^{\circ\prime}$, the ratio is very large in one direction and vanishingly small in the other. This means that for the less negative potentials, essentially all the electroactive solute is in the oxidized form (not surprising, since we started with the substance in this state). At the other extreme, there is essentially no solute in the oxidized state at the surface of the electrode.

TABLE 4-1
Necessary Ratio of Concentrations $C_{Zn^{++}}^s/C_{Zn(Hg)}^s$
at Electrode Surface for Various Applied Potentials (E)

E	$E - E^{\circ\prime}$	$C_{Zn^{++}}^s/C_{Zn(Hg)}^s$
0	1.00	6.08×10^{33}
-0.50	0.50	7.80×10^{16}
-0.90	0.10	2.39×10^3
-0.95	0.05	4.89×10^1
-0.99	0.01	2.18×10^0
-1.00	0.00	1.00
-1.01	-0.01	4.59×10^{-1}
-1.05	-0.05	2.05×10^{-2}
-1.10	-0.10	4.18×10^{-4}
-1.50	-0.50	1.28×10^{-17}
-2.00	-1.00	1.64×10^{-34}

The ratio of aqueous zinc ions to zinc atoms dissolved in mercury (as given by the table) determines the current that flows through the cell in the following way. At potentials more positive than about +100 mV relative to $E^{\circ\prime}$, the ratio is large, but here the mercury contains no zinc, so the requirement is already fulfilled and very little faradaic current is needed. On the other hand, as the potential becomes more negative, the ratio calls for a *large* concentration of zinc in the amalgam, and so a large current might be expected. Here, however, the current is limited by the rate of diffusion of Zn^{++} ions from the bulk of the solution to the electrode surface, and a plateau in the current-voltage curve results. At intermediate potentials, near $E^{\circ\prime}$, the current is determined by diffusion across a lesser gradient and has an intermediate value.

The preceding analysis is based on the assumption that at all times the rate of the electrode reaction is limited only by diffusion. This is equivalent to assuming that the electrochemical system is *reversible*.

In the case of an irreversible system, the shape of the curve will be modified somewhat, though the initial and final currents will not differ from the reversible case. At positive potentials, the curves coincide because so little chemical work is called for that the current required is easily available. At negative potentials, they again coincide because here sufficient overvoltage has been applied so that the electron-transfer reaction proceeds rapidly. In between these extremes, however, diffusion can supply reducible material to the surface faster than the electrons can be made available to react with it, and the curve becomes unsymmetrically drawn out, as sketched in Figure 4-6.

As noted previously, a large concentration of an inert *supporting electrolyte* must be present in addition to the electroactive species under study. This has several important effects. It increases the conductivity of the solution, hence diminishing IR voltage drops; it makes certain that the double layer is established in a reproducible manner; it maintains a constant ionic strength and thereby ensures that the activity coefficient of the species under study does not change as the charge-transfer reaction proceeds; and it swamps out the migration current that otherwise would be a variable perturbing effect.

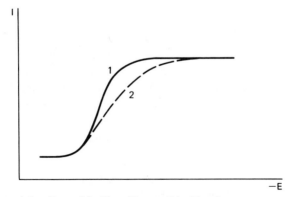

Figure 4-6.   Reversible (1) and irreversible (2) polarograms compared.

If no supporting electrolyte were present, migration would transport a considerable fraction of the ionic active species across the solution to the electrode. This would change the concentration profiles as defined by diffusion and thus alter the current. Such a change is not objectionable *per se*, since the current is often increased as a result, but it has the fundamental disadvantage that it is affected by inert species such as Na^+ or SO_4^- which would compete in carrying the current. Consequently, such species would constitute interferences. This mode of operation is generally not useful for analytical purposes.

THEORETICAL CONSIDERATIONS

The Ilkovič Equation

Since in a polarographic experiment, the current is controlled by the concentration of the electroactive species, it is desirable to establish a mathematical relation between these variables. To do this, it is expedient to deal first with the equation for diffusion to a stationary electrode.

If the electrode is planar and the system is diffusion controlled, the response to a large voltage step follows the Cottrell equation, previously introduced:

$$I = nFA(D/\pi t)^{1/2} C_{OX}^*$$ (4-3)

This equation is valid only for potentials that are sufficiently negative to ensure that the surface concentration $C^s = 0$. A more general equation can be obtained if we remove this restriction. In this case, the current is given [7] by:

$$I = nFA \left(\frac{D}{\pi t}\right)^{1/2} \left(\frac{1}{1 + \theta}\right) C_{OX}^*$$ (4-4)

in which

$$\theta = \exp\left\{\frac{RT}{nF} (E - E_{1/2})\right\} \cong C_{OX}^s / C_{RED}^s$$ (4-5)

The relation between the exponential and the ratio of concentrations can be derived by taking the logarithm of each side of the Nernst equation and substituting $E_{1/2}$ for $E^{o\prime}$. The exact significance of $E_{1/2}$ will be explained shortly.

We will now show how this equation can be modified to apply to the DME, in which the electrode area increases with time. It is assumed that the functional relation of Eq. (4-4) is retained except that the quantity A becomes a variable.

The area of a mercury drop under gravity control, as a function of time, can be shown by simple geometry to be:

$$A_{drop} = (4\pi)^{1/3} \left(\frac{3mt}{\rho}\right)^{2/3}$$ (4-6)

where m is defined as the rate of flow of mercury from the capillary, and ρ is its density. Inserting this value into Eq. (4-4) gives:

$$I = nF (4\pi)^{1/3} \left(\frac{3mt}{\rho}\right)^{2/3} D_{OX}^{1/2} (\pi t)^{-1/2} C_{OX}^* \left(\frac{1}{1 + \theta}\right) \qquad (4\text{-}7)$$

or

$$I = (\text{const}) \cdot nm^{2/3} t^{1/6} D_{OX}^{1/2} C_{OX}^* \left(\frac{1}{1 + \theta}\right) \qquad (4\text{-}8)$$

where the constant factor includes the denisty of mercury, $\rho = 13.6 \text{ g} \cdot \text{cm}^{-3}$, as well as the Faraday constant. A correction must be made to account for the contraction of the diffusion layer as the drop expands into the solution; this factor has been shown [8] to be $(\frac{7}{3})^{1/2}$. This gives the relation:

$$I = 7.06 \times 10^3 \cdot nm^{2/3} t^{1/6} D_{OX}^{1/2} C_{OX}^* \left(\frac{1}{1 + \theta}\right) \qquad (4\text{-}9)$$

Equation (4-9) gives I in amperes when m is in grams per second, t in seconds, D in centimeters-squared per second, and C in moles per liter. It describes the current during the lifetime of each drop, hence is inherently discontinuous with respect to both time and voltage. The equation must be applied separately for each drop, using the newly incremented voltage. This equation agrees with experimental results within a few percent.

Two limiting cases need to be treated: (1) where E is much more negative than $E_{1/2}$, so that we are situated on the plateau of the polarogram, and (2) the case where only the last instant of the lifetime of each drop is considered, and time t is replaced by a constant τ called the *drop time*.

At a sufficiently negative voltage, the value of θ becomes small, since it approximates the ratio C_{OX}^s/C_{RED}^s, which itself becomes very small, as seen in Table 4-1. In this case Eq. (4-9) reduces to the *Ilkovič equation* [9] :†

$$I = 7.06 \times 10^3 \cdot nm^{2/3} t^{1/6} D_{OX}^{1/2} C_{OX}^* \qquad (4\text{-}10)$$

The *diffusion current*, i_d, is the net current flowing just prior to the end of the drop-life, as shown in Figure 4-7. The drop time, τ, can be substituted for t to give:

$$i_d = 7.06 \times 10^3 \cdot nm^{2/3} \tau^{1/6} D_{OX}^{1/2} C_{OX}^* \qquad (4\text{-}11)$$

†The numerical multiplier is often given as 706, which gives the current in microamps if m is taken in milligrams per second and the concentration in millimoles per liter. Older instrumentation, now obsolete, used a highly damped recording mechanism that gave, in effect, the *average* current. To account for this, Eq. (4-10) must be integrated and divided by the drop time, to give:

$$\overline{I} = 605 \, nm^{2/3} \tau^{1/6} D_{OX}^{1/2} C_{OX}^* \qquad (4\text{-}10a)$$

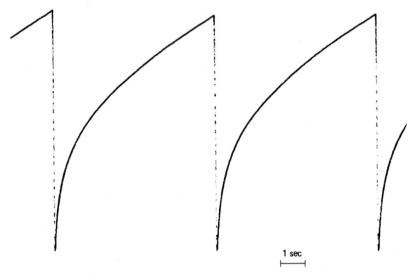

Figure 4-7. Polarographic trace covering two drop lives. Natural drop time, about 6.5 s; paper speed, 1 cm/s. (Recorded on a Sargent-Welch Polarograph.)

These equations demonstrate a linear relation between the diffusion current and the concentration, a property which is exploited in analytical measurements.

The order of magnitude of the diffusion current as a function of n and C_{OX}^* may be seen by taking as typical values, $m = 0.002$ g $\cdot$ s^{-1}, $\tau = 4$ s, and $D_{OX} = 6 \times 10^{-6}$ cm^2 s^{-1}, giving the results shown in Table 4-2.

It is noteworthy that the temperature does not appear explicitly in the Cottrell and Ilkovič equations. It does, however, have an effect, primarily through the temperature dependence of the diffusion coefficient and of the physical properties of mercury. The net effect on the diffusion current is to increase its value by 1 or 2 percent per degree rise of temperature in the absence of kinetic complications. For this reason, it is advisable to control the temperature of a polarographic cell to within about 0.5°C if quantitative work is contemplated.

The Ilkovič equation can be made more precise by the addition of one or more terms (p. 114 of Reference 4). For analytical purposes this is not necessary, how-

TABLE 4-2
Typical Diffusion Currents in Microamperes
(Calculated)

Concentration (moles/liter)	n		
	1	2	3
10^{-3}	8.5	17	25
10^{-4}	0.85	1.7	2.5
10^{-5}	0.085	0.17	0.25

ever, since the utility of the equation is primarily dimensional, showing the relative importance of its several variables. In practice, the relation between concentration and diffusion current should always be determined experimentally by making measurements on standard solutions. For work with the SMDE, the Cottrell equation applies directly.

The Heyrovský–Ilkovič Equation

We will now derive an equation for the shape of the curve of faradaic current plotted against applied potential, which is approximated by the sampled polarogram. This is tantamount to taking the locus of all points for which $t = \tau$. Note that now the current is independent of time.

Substitution of the value of i_d from Eq. (4-11) into Eq. (4-9) gives (with replacement of t by τ):

$$I = i_d \left(\frac{1}{1 + \theta}\right) \tag{4-12}$$

which can be solved for θ to give:

$$\theta = \frac{i_d - I}{I} \tag{4-13}$$

From Eq. (4-5), by taking logarithms, we can write:

$$E = E_{1/2} + \frac{RT}{nF} \ln\left(\frac{i_d - I}{I}\right) \tag{4-14}$$

This equation, first derived by Heyrovský and Ilkovič [10], gives the shape of the current-voltage curve, which is centro-symmetric about $E_{1/2}$. The relationship between i_d and I is evident in Figure 4-8.

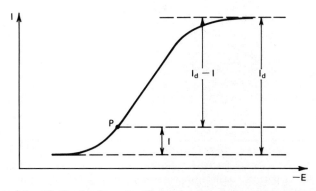

Figure 4-8. Idealized polarogram, illustrating the relations between I and i_d.

The half-wave potential, $E_{1/2}$, has two definitions. Mathematically, it is defined as

$$E_{1/2} = E^{\circ\prime} + \frac{RT}{nF} \ln \left(\frac{D_{RED}}{D_{OX}} \right)^{1/2} \qquad (4\text{-}15)$$

whereas an operational definition describes it as the potential at which $I = i_d/2$. The two definitions coincide for reversible reactions.

The diffusion coefficients seldom differ by more than a factor of 2, corresponding to less than 20 mV. Hence, $E_{1/2}$ is usually not far different from $E^{\circ\prime}$, either for the case where the oxidized species is a hydrated cation and the reduced form is soluble in mercury, or for the case in which both oxidized and reduced species are soluble in water. If one or both water-soluble species are complexed by some component of the solution, or if kinetic complications appear, then $E_{1/2}$ may differ considerably from $E^{\circ\prime}$.

If the function $-\log [(i_d - I)/I]$ is plotted against $-E$, a straight line should result with slope equal to 0.0592/n volts (at 25°C), crossing zero at $E_{1/2}$ (Figure 4-9). This provides a convenient method of determining both n and $E_{1/2}$ for reversible systems.

Irreversible Systems

The equations presented in the preceding pages apply to ideal, reversible reactions. We now turn to irreversible systems for which some of the equations must be modified.

The basic distinction between the two classes lies in the magnitude of the heterogeneous rate constant, k°, as defined in Chapter 2, corresponding to the reaction:

$$OX + ne^- = RED \qquad (4\text{-}16)$$

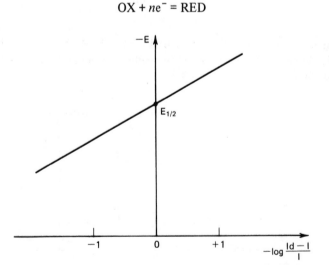

Figure 4-9. Plot of potential against the negative log of $[(i_d - I)/I]$, showing how to determine the half-wave potential.

A convenient criterion for reversibility [11] is the relative values of the quantities $2(D/\pi\tau)^{1/2}$ and $k°$, both of which have the dimensions of cm/s. The former can be considered to be a diffusion rate constant, k_d, while $k°$ is the charge-transfer rate constant. We can state the conditions for reversibility† as:

$$k° \gg k_d \quad \text{(reversible)} \tag{4-17}$$

$$k° \ll k_d \quad \text{(irreversible)} \tag{4-18}$$

In the case of polarography, we can take as typical values $D = 5 \times 10^{-6}$ cm^2 s^{-1}, $\tau = 5$ s, which gives for k_d a fairly constant value, about 1×10^{-3} cm · s^{-1}. As a rule, for a truly reversible reaction, $k° > 2 \times 10^{-2}$, while a reaction for which $k° < 2 \times 10^{-5}$ is considered to be totally irreversible.

In an electrochemical process, the tendency for a reaction to occur is determined by the applied potential E, and therefore the reference potential must be specified. For present purposes, it is most convenient to take the formal potential $E^{°\prime}$ as this reference. The following equations will be consistent with this standard.

As previously stated, the Cottrell equation is valid for both reversible and irreversible reactions because the potential step on which it is based extends sufficiently negative that all OX molecules reaching the electrode are immediately reduced, regardless of the magnitude of the rate constant. For a voltage step of arbitrary amplitude, the following relation can be obtained from Eq. (4-4):

$$I = \frac{nFAD_{OX}^{1/2}}{(\pi t)^{1/2} \left[\exp\left\{ \frac{nF}{RT} (E - E^{°\prime}) \right\} + 1 \right]} \cdot C_{OX}^* \tag{4-19}$$

in which it is assumed that $D_{OX} = D_{RED}$, so that $E_{1/2}$ can be replaced by $E^{°\prime}$. Equation (4-19) reduces to Eq. (4-3) for sufficiently negative potentials. This general relation, however, assumes reversibility, since it utilizes the Nernst equation in its derivation.

For an irreversible reaction, the comparable expression is (p 235 of [4]):

$$I = \frac{nFAk°}{\exp\left\{ \frac{\alpha nF}{RT} (E - E^{°\prime}) \right\}} \cdot C_{OX}^* \tag{4-20}$$

where α is the transfer coefficient. Note that $(D_{OX}/\pi t)^{1/2}$ no longer appears, and $k°$ takes its place, indicating that the process is now under kinetic rather than diffusion control. For potentials sufficiently negative, the denominator becomes very small (for $E - E^{°\prime} = -1$ V, when $n = 2$, and $\alpha = 0.5$, the exponential is of the order of 10^{-17}), and this predicts an impossibly large current. The fact that the current does not really go this high is evidence that diffusion must take over control if the potential step is large enough.

---

†Compare these definitions with the treatment of reversibility based on exchange currents, in Chapter 2.

Note that the Ilkovič equation [Eq. (4-11)], relating the diffusion current to concentration, is equally valid for reversible and irreversible processes, since the measurement of i_d is made under conditions that ensure diffusion control. Both cases are analytically useful.

The derivation leading to the Heyrovský relation [Eq. (4-14)] is no longer valid for an irreversible reaction, as one might expect from the difference in curve shapes (Figure 4-6). The derivation is too involved to justify repeating here. The final equation (p. 240 of Reference 4) is:

$$E = E_{1/2} + \frac{0.0235}{\alpha n} \ln \left(\frac{i_d - I}{I} \right) \tag{4-21}$$

where

$$E_{1/2} = E^{\circ\prime} + \frac{RT}{\alpha n F} \ln \left[k^{\circ} \left(\frac{12\tau}{7 D_{OX}} \right)^{1/2} \right] \tag{4-22}$$

Note that $E_{1/2}$ is still the potential at which the current equals half the diffusion current, but its numerical significance is different. In this irreversible case, $E_{1/2}$ changes also with the drop time, τ.

Meites (p. 224 of Reference 4) has suggested a method of testing for reversibility by measurement (on the polarogram) of $E_{1/4}$ and $E_{3/4}$, the potentials at which the corrected current is, respectively, 1/4 and 3/4, of i_d. For a reversible reaction, one can show that:

$$E_{1/4} = E_{1/2} - \frac{0.0592}{n} \log (1/3) \tag{4-23}$$

and

$$E_{3/4} = E_{1/2} - \frac{0.0592}{n} \log 3 \tag{4-24}$$

Subtraction gives:

$$E_{3/4} - E_{1/4} = - \frac{0.0565}{n} \tag{4-25}$$

For an irreversible system, $E_{3/4} - E_{1/4}$ is generally more negative than this. This method can be used to determine n if the reaction is reversible, or as a test for irreversibility if n is known, whereas the method of Figure 4-9 is only valid for reversible systems.

Anodic Oxidations

So far we have considered only cathodic reductions, but the DME can also act as an anode. Figure 4-10 shows the polarograms that result from a mixture of oxi-

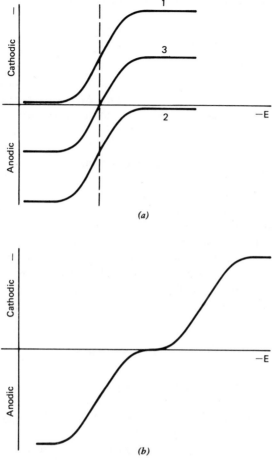

Figure 4-10. Composite anodic and cathodic curves. (a) A reversible system, showing the pure components plus the mixture. (b) An irreversible system; the curve does not cross the zero-current axis abruptly, as in (a), but becomes tangent to it, because of the separation of half-wave potentials of the oxidized and reduced forms.

dized and reduced forms of the same element. Curve 1 is the usual reduction curve obtained when only the oxidized form is present. If only the reductant is present in the solution, the anodic curve 2 is seen, and a solution containing equal quantities of the two oxidation states gives a response such as curve 3. The fact that all three curves show the same $E_{1/2}$ indicates that the redox process in this medium is reversible. The ratio of the diffusion currents i_d for the cathodic and anodic portions (for equal concentrations) depend on the diffusion coefficients:

$$\frac{i_{d(cath)}}{i_{d(anod)}} = \left(\frac{D_{OX}}{D_{RED}}\right)^{1/2}$$

(4-26)

A relation analogous to Eq. (4-14) can be derived to cover the present case:

$$E = E_{1/2} + \frac{RT}{nF} \ln\left(\frac{i_{d(cath)} - I}{I - i_{d(anod)}}\right) \qquad (4\text{-}27)$$

Clearly, if the reduced form is absent, $i_{d(anod)}$ will be zero and Eq. (4-27) will reduce to (4-14). An irreversible system with equal concentrations of oxidized and reduced species present will give a composite curve such as that in Figure 4-10b.

Polarographic Maxima

In a typical DC polarographic experiment, a small amount of a surface-active agent is frequently needed. This is to prevent the formation of distortions called *maxima*.

A large variety of organic compounds have been found effective as maximum suppressors. By far the most widely used today is a nonionic detergent known as Triton X-100.† Others that have been used include gelatine and methyl red. It is important not to use too much suppressor, as this may cause worse trouble than it cures. About 0.002 percent of Triton X-100 is usually satisfactory, but one should use only as little as will be effective.

INSTRUMENTATION

A DC polarograph must contain a unit for producing a voltage ramp to be applied to the cell, a potentiostat, and a current-measuring device. Modern instruments are automatically recording and include a ramp generator that is synchronized with the recorder.

Figure 4-11 gives block diagrams of two widely used versions of polarographic instrumentation.‡ The unit marked "ramp generator" consists of either an electronic integrator that gives a potential $E = k \int dt = kt$, or a step function generator that produces a staircase approximation to a ramp. In either case, its output is an increasing voltage, adjustable in both range and slope. This ramp voltage is tracked by the potentiostat (amplifier No. 1 or 3) and impressed upon the solution as monitored by the reference electrode. In circuit (a) the DME is directly connected to ground, and the current is measured by the voltage drop across resistor R_1 in series with the auxiliary electrode, as sensed by the differential amplifier, No. 2. In the alternative circuit at (b), the DME is held at *virtual* ground by the action of amplifier No. 4, but the cell current is forced to flow through

---

†Rohm and Haas Company, Philadelphia.

‡Readers who are not familiar with the properties and applications of operations amplifiers (op amps) and other electronic circuits are referred to Chapter 17. It must be noted that the word "Polarograph" is the registered trademark of Sargent-Welch Scientific Company, Skokie, Illinois.

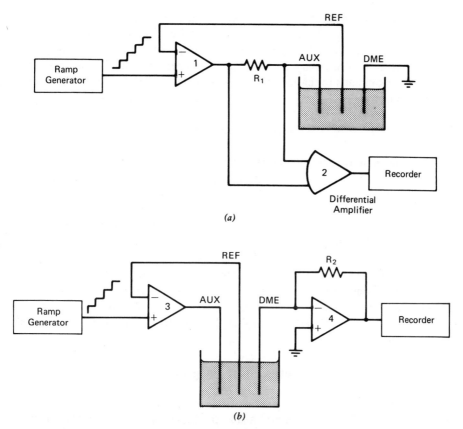

Figure 4-11. Simplified schematic diagrams of two common polarographic circuits. In (*a*) the DME is directly grounded, thereby reducing electrical noise pick-up. Circuit (*b*) employs a current-to-voltage converter (amplifier #4).

resistor R_2 to give to the recorder a proportionate output voltage. Circuit (*a*) is less prone to noise pickup because the mercury handling system can be directly grounded.

Sampling

As we have seen, polarograms are simplified by provision of a method of recording the current during only a brief interval just prior to the fall of each drop. This can be accomplished with an electronic sampling circuit that, in effect, connects the current measuring amplifier (No. 2 or 4 in Figure 4-11) to the recorder only during the desired time intervals.

A precise timing circuit is needed to ensure that the sampling window is correctly placed in the life of each drop. There are two possible ways to identify the birth of a drop: a circuit can be arranged to detect the fall of the previous drop and start the timing sequence [12], or an electromechanical drop-knocker can be provided to cause the mercury to fall at preselected intervals. The latter method is usually preferred, since it ensures an adjustable, uniform drop time throughout an experiment. Timing circuits for either option are easily designed with standard electronic components.

Resistance Compensation

Under some circumstances, the resistance of the solution may become significant. This tends to increase the uncertainty of the actual potential across the interface, since the drop in voltage produced by the resistance of the solution changes with the current that is being measured. This is especially difficult to handle with a two-electrode cell. The three-electrode form permits the location of the liquid junction from the reference electrode to be strategically placed, so that nearly all of the IR drop is eliminated. Even with a three-electrode cell, however, uncompensated resistance is apt to become objectionably large in nonaqueous solvents or with low concentrations of supporting electrolyte. The uncompensated IR drop can be corrected for by subtracting a voltage proportional to the current from the applied potential. This can be done within the potentiostat circuitry, and is called *compensation by positive feedback*.

Charging-Current Compensation

The current that flows through a polarographic cell in the absence of an electroactive species, the *residual current*, is due to the charging of the double layer at the DME. It is given by the product $C_{dl}(E - E_{PZC})(dA/dt)$. The derivative dA/dt can be seen from Eq. (4-6) to be a function of $t^{-1/3}$. If sampling is used, dA/dt is a constant. The residual-current curve approximates two straight lines intersecting at the PZC, the more negative segment having a lesser slope than the more positive (because of the difference in the double-layer capacitance in these two regions).

The slope of the residual current causes difficulty with dilute solutions because of the necessary increase in amplification. The situation can be improved by a procedure known as *linear baseline compensation*, which consists of adding to the output an opposing voltage ramp just sufficient to return the residual current trace to an approximately horizontal line. The compensation can never be perfect, but it can nevertheless be very helpful. Figure 4-12 shows an example of its use.

A polarograph with all the foregoing features is diagrammed in Figure 4-13.

Cell Design

Much of the early work in polarography was carried out with a simple two-electrode cell in which a large pool of mercury served as combined auxiliary and reference electrode (Figure 4-14*a*). It was necessary for the supporting electrolyte

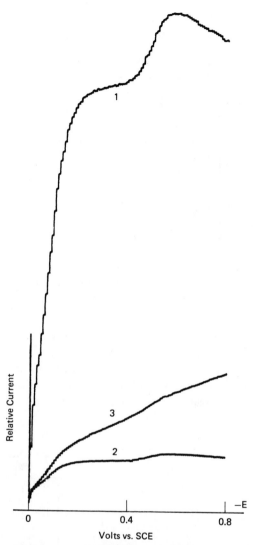

Figure 4-12. Sampled DC polarograms of 0.64 ppm thallium in 0.5 M KCl. Curve 3: without compensation; the wave at $E_{1/2} = 0.45$ V is hardly discernable. Curve 2: same solution, but with linear compensation; the reduced background slope makes it possible to increase the gain 10 times to give Curve 1, from which a fairly precise measurement of the wave height can be made. (Recorded on a Sargent-Welch Polarograph.)

to contain an "anodic depolarizer," such as chloride or other ion that forms an insoluble salt with mercury, to provide a stable reference potential.

The two-electrode design reached its zenith with the H-cell (Figure 4-14b), wherein the reference electrode is permanently attached by a connecting salt bridge closed with a glass frit and filled with a KCl-containing plug of agar gel.

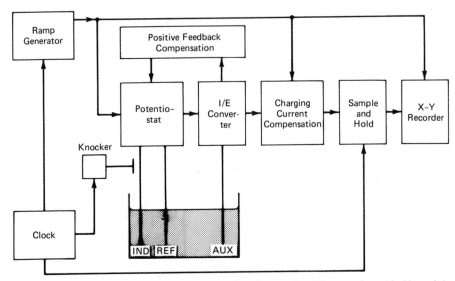

Figure 4-13. Block diagram of a modern DC polarograph. The sample-and-hold module retains the value of the current for each drop until up-dated with information from the next drop.

Since the general introduction of the three-electrode cell concept, the H-cell has become obsolescent. More rugged and less expensive is a simple beaker (100 mL is a convenient size) provided with a cover of Teflon or other inert plastic (Figure 4-14c). The cover is pierced with holes to accommodate the DME capillary, a reference electrode, a platinum button or wire auxiliary electrode, and a gas-

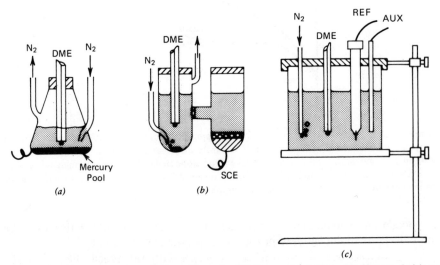

Figure 4-14. Representative polarographic cells. (a) Heyrovský design. (b) Lingane–Laitinen H-cell. (c) Suggested three-electrode cell with supporting cover; see also the frontispiece.

inlet tube. Whatever cell design is adopted, provision should be made for temperature control.

The Capillary System

A typical glass capillary through which mercury is delivered to form a DME is about 50 μm internal diameter and 15 cm in length.† It is important that it be supported vertically and that the lower end be cut squarely. The capillary is connected to a reservoir by flexible tubing. The capillary for use with the SMDE is shorter and with larger bore to allow a drop to form quickly.

A capillary, if properly cared for, should last for many months. A basic precaution that must be taken is to prevent any dissolved solid matter from entering the capillary opening. Another precaution is never to allow the potential to become so positive as to permit the anodic oxidation of mercury. Some of the Hg_2Cl_2 or other salt so formed is certain to get into the capillary and render it useless. The DME is quite sensitive to vibration, and the location of the instrument should be selected carefully with this in mind.

EXPERIMENTAL METHODOLOGY

Supporting Electrolytes and Half-Wave Potentials

A great variety of materials have been used as supporting electrolytes. Some are listed in Table 4-3 [15]. The choice is sometimes not at all easy, particularly if more than one active species is present. Table 4-4 lists a selection of half-wave potentials in the electrolytes mentioned. Study of this table suggests that species which interfere in one medium can often be resolved in another. For example, Sn(IV) and Zn(II) are reported to have identical $E_{1/2}$ values in acetate, whereas they are widely separated in H_3PO_4. As another example, suppose that two waves are obtained in HCl, with $E_{1/2}$ about -0.43 and -0.65 V. This combination might be ascribed to the double wave of As(III) or just as well to a mixture of Pb(II) and Cd(II). A shift to an acetate medium would clarify the situation.

It is possible to gain much information about complex formation, including equilibrium constants, from polarographic studies. The $E_{1/2}$ for the reversible reduction of a complex to a metal soluble in mercury is displaced toward more negative potentials as compared to the uncomplexed hydrated metal ion. If both oxidized and reduced species are water-soluble and form complexes, the situation becomes

---

†Efforts have been made to fabricate capillaries of some material that is resistant to such corrosive electrolytes as HF and concentrated alkali. Raaen [13] has reported a Teflon capillary that can give excellent results but is difficult to prepare. More recently Ménard and LeBlond-Routhier [14] have described a more easily made capillary with a polyethylene tip that is serviceable for these applications.

TABLE 4-3
Supporting Electrolytes for Polarography[a]

For Aqueous Solutions

1.	Neutral	$NaClO_4$ or KNO_3
2.	Acetate	$2\ M$ HOAc + $2\ M$ NH_4OAc
3.	Ammonia	$1\ M$ NH_3 + $1\ M$ NH_4Cl
4.	Chloride	$0.1\ M$ KCl or NaCl
5.	HCl	$1\ M$ HCl
6.	Citrate	Citric acid + Ammonium citrate, pH = 4
7.	Citrate	Same, pH = 6
8.	EDTA	$0.1\ M$ EDTA, pH = 7
9.	Hydroxide	$1\ M$ KOH or NaOH
10.	Oxalate	$0.25\ M$ Oxalic acid + $(NH_4)_2$-Oxalate, pH = 4
11.	Phosphate	$7.3\ M$ H_3PO_4

For Nonaqueous Solutions

LiCl, Me_4NClO_4, Bu_4NI, $LiClO_4$, according to solubilities.

[a]Selected from a much larger list in Reference 15.

more involved, and the shift of potential can be in either direction. In irreversible systems, the effect of complex formation may also be quite diverse.

It is because of the great variation of half-wave potentials from one electrolyte to another that polarography is seldom a technique of choice for *qualitative* analysis. It can, of course, serve negatively to establish the *absence* of some element. A more general qualitative scheme for the elements would necessitate running the same sample in various supporting electrolytes and correlating the data so obtained. For more than three or four ions, this would become too cumbersome. Qualitative identification of organic compounds is likely to be even less productive.

Resolution

Resolution in this context refers to the needed separation of two waves. The basic requirement is that the half-wave potentials be sufficiently far apart that the plateau between them will become nearly parallel to the baseline. Generally, two reversible waves of equal height and the same *n*-value may be considered resolved if their half-wave potentials are separated by $300/n$ mV, but this is only a guide rather than a firm criterion. For irreversible reductions, the required separation is considerably greater.†

†Ružić and Branica [16] have presented a graphical method by which a polarogram consisting of two overlapping waves can be analyzed. The individual half-wave potentials and the ratio of the two diffusion currents can be obtained. The values of *n* for the case of reversible reactions (or of α*n* for totally irreversible reactions) can also be found. The method is rather cumbersome, and is not applicable to quasi-reversible systems.

TABLE 4-4

Half-Wave Potentials of Elements in Aqueous Solution [15]

(vs. SCE)

Supporting Electrolytes	As^{III}	Cd^{II}	Co^{II}	Cr^{IV}	Fe^{II}	Fe^{III}	In^{III}	Pb^{II}	Sn^{II}	Tl^{I}	Zn^{II}
1	-0.07 -1.00	—	—	—	-1.46*	>0	—	-0.38	(+0.14)* -0.43	—	-1.00
2	-0.92	-0.65	-1.19*	—	NR	>0	-0.71	-0.50	(-0.16) -0.62	-0.47	-1.10*
3	-1.41* -1.63	-0.81	-1.29*	-0.20 -0.60	(-0.34)* -1.49	ppt	—	ppt	—	-0.48	-0.20 -1.60
4	—	-0.60	-1.20*	-0.30* -1.00 -1.80	-1.30*	—	-0.56	-0.40	—	-0.46	-1.00
5	-0.43* -0.67	-0.64	NR	—	NR	>0	-0.56	-0.44	(-0.10) -0.47*	-0.48	NR
6	-0.75	-0.59	NR	—	(-0.05)	+0.05	—	-0.43	(-0.21)* -0.54	-0.45	-1.04*
7	-1.46*	-0.70	NR	>0 -0.36	(-0.18)	-0.23*	NR	-0.56	(-0.41)* -0.66	-0.51	-1.37*
8	—	-1.27	NR	—	—	-1.15	-1.08	-1.37	-1.26	-0.48	NR
9	(-0.26)	-0.78*	-1.46*	-0.85*	(-0.90)	—	-1.09	-0.76	(-0.73)* -1.22	-0.48	-1.53
10	-0.79	-0.63	—	—	(-0.23)	-0.23	—	0.50	-0.70	-0.46	—
11	-0.46 -0.71	-0.77	-1.20*	>0	NR	+0.06	—	0.53	-0.58*	-0.63	-1.13*

Notes: * Irreversible; NR not reduced; — data not available; ( ) anodic wave; two or more entries, multiple wave; supporting electrolytes; see Table 4-2.

It must be realized that the number of electroactive species that can be determined in a given supporting electrolyte is severely limited by this resolution requirement. At equal concentrations, for $n = 2$, the maximum number of waves theoretically resolvable is $2/0.15 = 13$, where the available range is taken as 0 to -2 V. Of course this number cannot be attained in practice because half-wave potentials are not evenly distributed over the range.†

The separation necessary for adequate resolution depends also on the relative heights of the waves. This effect is likely to be more important where two adjacent waves are greatly different in height; the smaller wave is likely to be swamped out by the larger.

Precision, Accuracy, and Sensitivity

As in any analytical method, it is desirable to estimate the precision that it is reasonable to expect. Fisher [17] has considered this carefully for the case of DC polarography and concludes that quantitative measurement of a $5 \times 10^{-3} M$ solution of Cd^{++} can easily reach a precision of 2 percent of the amount present. This can be improved by perhaps an order of magnitude with special precautions.

The accuracy of a determination, of course, depends not only on the precision (reproducibility), but also on the accuracy with which the concentrations of standards are known.

The sensitivity of the polarographic method is limited by the magnitude of the charging current. The practical limit of useful measurements is reached with about $10^{-4} M$ solutions for the conventional (DC) method.

Scope of Applicability

Polarographic analysis is applicable to many kinds of substances. The principal requirements are that a substance be soluble in an ionizing solvent and that it be oxidizable or reducible at a potential within the range accessible to the DME, namely about $+0.4$ to beyond -2 V (in aqueous solution, vs. SCE). The electroactive substance may be an ion of either charge or a neutral species.

The majority of polarographic analyses involve reductions, and nearly any metal, except perhaps the most active, can be reduced from its salts or complexes, either to the metal or to the lower oxidation state. Double waves are often seen with species such as chromate, that can be reduced in two steps.

Anodic oxidations that are observable polarographically include metallic species that exist in a lower oxidation state in solution. Occasionally a useful anodic wave can be obtained that involves oxidation of mercury. For example, traces of chloride can be determined by the reaction:

$$2Hg + 2Cl^- - 2e^- = Hg_2Cl_2$$

---

†It may be instructive to compare this limited number of "channels" with a comparable quantity for atomic absorption: a typical bandwidth for an AA peak may be taken as 10^{-4} nm, which would permit 3×10^6 channels in a range of 300 nm.

Anodic oxidation of metals from an amalgam are treated in some details in Chapter 12.

Organic Applications

A great variety of organic compounds can be reduced at the DME. A representative list is given in Table 4-5 [18]. To be reducible under polarographic conditions, a compound must possess one or more polar or unsaturated linkages. The half-reactions often require H^+ ions for completion. As an example [19], the two-step reduction of α-aminoacetophenone follows the mechanism:

TABLE 4-5
Some Organic Compounds Reducible at the DME[a]

Class	Example	Medium	$E_{1/2}$	n
Carboxylic Acids	Acetic acid	Et_4NClO_4/MeCN	-2.3	
	Fumaric acid	NH_3 buffer/10% EtOH	-1.57	
	Maleic acid	NH_3 buffer/10% EtOH	-1.35	
	Oxalic acid	Et_4NClO_4/MeCN	-1.6	
Esters	Ethyl acrylate	Me_4NI/30% EtOH	-1.82	
	Diphenyl phthalate	Me_4NCl/75% EtOH	-1.65	
Carbonyl Compounds	Acetaldehyde	LiOH, aq.	-1.93	
	Acetone	Bu_4NCl, BuNOH/80% EtOH	-2.53	2
	Me_2-glyoxime	HCl/15% EtOH	-0.81	8
Halogen Compounds	Allyl bromide	Li_3Cit/50% dioxane	-1.18	2
	Bromobenzene	Et_4NBr/DMF	-2.24	2
	2-Cl-cyclohexanone	KCl, aq.	-1.45	
Heterocycles	2,2'-bipyridine	Acetate buffer, KCl, aq.	-1.14	1
	Coumarin	Phosphate buffer, pH 6.8	-1.53	2
	Pyrimidine	Phosphate buffer, pH 6.8	-1.30	2
	Quinoline	Me_4NOH/50% EtOH	-1.50	2
Unsat'd Hydrocarbons	Anthracene	Bu_4NI/75% dioxane	-1.94	2
	Naphthalene	Bu_4NI/75% dioxane	-2.49	2
	Phenanthrene	Bu_4NI/75% dioxane	-2.44	2
	Styrene	Bu_4NI75% dioxane	-2.34	2
Nitro Compounds	Azobenzene	NaCl, KCl/30% MeOH	-0.81	2
and related	Ethyl nitrate	LiCl/10% EtOH	-0.82	2
	Nitrobenzene	Acetate buffer/60% EtOH	-0.43	
	Trinitroglycerol	Me_4NCl/75% EtOH	-0.70	
Sulfur Compounds	Cystine	NH_3 buffer/pH9.2	-1.3	
	Diphenyl-sulfoxide	Me_4NBr/50% EtOH	-2.07	2
	Thiourea	H_2SO_4, aq.	(+0.04)	
	(anodic wave; forms ppt with Hg)			

[a]Note: Reference electrodes must be selected with care, to be compatible with organic solvents.

$$\phi - CO - CH_2 - NR_3 \xrightarrow{H^+, 2e^-} \phi - CO - CH_3 \xrightarrow{2H^+, 2e^-} \phi - CHOH - CH_3$$

This was observed in a pH 7 phosphate buffer, so the required hydrogen ion was merely that available from the dissociation of solvent water.

Organic applications have been reviewed extensively by Zuman [19, 20] and by Elving [21].

Solubility is, of course, a source of difficulty with organic compounds. Many materials usually regarded as insoluble in water are, nevertheless sufficiently soluble to be measurable polarographically. Others are easily accessible in mixed solvents, such as 50 percent aqueous alcohol or dioxane, but many require anhydrous solvents such as acetonitrile or dimethylformamide. In aqueous solvents, a number of the familiar salts or acids are sufficiently soluble and ionized to serve as supporting electrolytes. In anhydrous media, tetra-alkylammonium salts are often the only suitable materials for this purpose.

The choice of reference electrodes is another problem with nonaqueous solvents, whether they are truly anhydrous or not. Frequently, a conventional calomel or silver chloride is selected with a liquid junction between aqueous KCl solution and the prevailing solvent. Such an electrode may serve empirically as a reference, but the expected large and unknown junction potential precludes accurate and reproducible data on half-wave potentials.

Various reference electrodes have been described for use in specific solvents. In many, a silver wire immersed in a solution of Ag^+ ions in the organic solvent will suffice. McMasters et al. [22] prepared a stable electrode for use in dimethylsulf-oxide, consisting of a saturated solution of zinc perchlorate in contact with a saturated zinc amalgam. Recent work by Peerce and Bard [23] has shown that a platinum electrode coated with a polymer of vinylferrocene serves as a good reference in acetonitrile, and suggests that polymer electrodes of other composition may be similarly useful. Other reference electrodes are described by Ives and Janz [24] and by Butler [25]. An interesting discussion of voltammetry in nonaqueous solvents can be found in Reference 26.

REFERENCES

1. J. Heyrovský, *Rec. Trav. Chim.*, 1925, *44*, 488.

2. J. Heyrovský and M. Shikata, *Rec. Trav. Chim.*, 1925, *44*, 496.

3. G. W. Ewing and J. E. Nelson, *J. Chem. Educ.*, 1969, *46*, 292.

4. L. Meites, "Polarographic Techniques," 2nd ed., Wiley, New York, 1965.

5. W. M. Peterson, *Amer. Lab.*, 1979, *11(12)*, 69.

6. A. M. Bond, *J. Electroanal. Chem.*, 1981, *118*, 381.

7. E. R. Brown and R. F. Large, in "Physical Methods of Chemistry," (A. Weissberger and B. W. Rossiter, Eds.), Wiley-Interscience, New York, 1971; Part IIA, p. 442.

8. D. MacGillavry and E. K. Rideal, *Rec. Trav. Chim.*, 1937, *56*, 1012.

9. D. Ilkovič, *Collect. Czech. Chem. Commun.*, 1934, *6*, 498.

10. J. Heyrovský and D. Ilkovič, *Collect. Czech. Chem. Commun.*, 1935, *7*, 198.

11. P. Delahay, in "Advances in Electrochemistry and Electrochemical Engineering" (P. Delahay, Ed.), Wiley, New York, **1961**, Vol. 1, p. 233.

12. P. D. Tyma, M. J. Weaver and C. G. Enke, *Anal. Chem.*, **1979**, *51*, 2300.

13. H. P. Raaen, *Anal. Chem.*, **1965**, *37*, 1355.

14. H. Ménard and F. LeBlond-Routhier, *Anal. Chem.*, **1978**, *50*, 687.

15. L. Meites, in "Handbook of Analytical Chemistry" (L. Meites, Ed.), McGraw-Hill, New York, **1963**, p. 5-98.

16. I. Ružić and M. Branica, *J. Electroanal. Chem.*, **1969**, *22*, 243.

17. D. J. Fisher, in "Electroanalytical Chemistry" (H. W. Nürnberg, Ed.), Wiley, New York, **1974**, p. 93ff.

18. P. Kabasakalian and J. H. McGlotten, in "Handbook of Analytical Chemistry" (L. Meites, Ed.), McGraw-Hill, New York, **1963**, p. 5-104.

19. P. Zuman, "Organic Polarographic Analysis," Pergamon Press, London, and Macmillan, New York, **1964**, p. 84ff.

20. P. Zuman, "Substituent Effects in Organic Polarography," Plenum Press, New York, **1967**.

21. P. J. Elving, in "Electroanalytical Chemistry," (H. W. Nürnberg, Ed.), Wiley, New York, **1974**, Chapter 3.

22. D. L. McMasters, R. B. Dunlap, J. R. Kuempel, L. W. Kreider and T. R. Shearer, *Anal. Chem.*, **1967**, *39*, 103.

23. P. J. Peerce and A. J. Bard, *J. Electroanal. Chem.*, **1980**, *108*, 121.

24. D.J.G. Ives and G. J. Janz, "Reference Electrodes," Academic Press, New York, **1961**.

25. J. N. Butler, in "Advances in Electrochemistry and Electrochemical Engineering" (P. Delahay and C. W. Tobias, Eds.), Wiley-Interscience, New York, **1970**, Vol. 7, p. 77ff.

26. J. Badoz-Lambling and G. Cauquis, in "Electroanalytical Chemistry" (H. W. Nürnberg, Ed.), Wiley, New York, **1974**, Chapter 5.

Chapter 5

VOLTAMMETRY:
II. PULSE AND SQUARE-WAVE POLAROGRAPHY

Many attempts have been made throughout the years to improve the sensitivity of the polarographic method. The most successful of these depend on some type of modulation, an approach that provides *two* qualitatively different parameters that can be measured, so that the effects of the charging current can be eliminated between them.

One of these methods is *pulse polarography*, in which a pulse of a few tens of millivolts is impressed on the staircase voltage toward the end of each drop-life. Two current measurements are taken, one just before the pulse, and another during the pulse, as shown in Figure 5-1. The *difference* between the current at B and that at A for each step, plotted against $-E$, gives a differential polarogram, essentially free from the effects of charging current. Pulse polarography and closely related methods are treated in this chapter.

Another technique, *AC polarography*, calls for the superimposition of a small amplitude AC voltage on the staircase ramp (AC modulation). The signal is taken as the AC component of the resulting current. Since alternating current is characterized by two parameters, its amplitude and phase, careful attention to phase relations permits the determination of faradaic current free of charging current. This will be explored in the following chapter.

Both methods achieve a marked improvement of perhaps a thousandfold in sensitivity, compared to conventional (DC) polarography.

A BASIC EXPERIMENT

A possible measurement scheme is shown in Figure 5-2. Just as in sampled DC polarography, the instrument must contain some device that causes the drop to

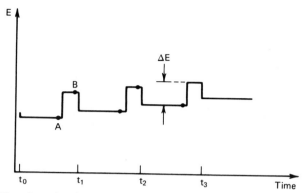

Figure 5-1. Wave form for the applied voltage in differential pulse polarography. The drop is knocked off at times $t_0, t_1, t_2, \ldots.$ Measurements are taken at points, marked A and B.

fall at a known moment. The sampler takes instantaneous current readings at times A and B to be fed into a difference circuit that generates a voltage proportional to $(I_b - I_a)$, the desired output. In Figure 5-3, it can be seen that for each drop, the current up to point A is the same as in DC polarography, while at B the current contains in addition the effect of the voltage step. By taking the difference, the DC polarogram is subtracted out, and one obtains exclusively the response to the pulse. This response consists of two components: (1) the current necessary to charge the double layer from E to the new potential $E + \Delta E$, and (2) the portion of the faradaic current resulting from the pulse. As seen in Figure 5-3, the charging current decays faster than the faradaic current, so that a judicious choice of point B permits almost complete elimination of the charging current. This represents the major strength of pulse polarography.

The fact that a difference is taken results in a derivative-like form for the signal,

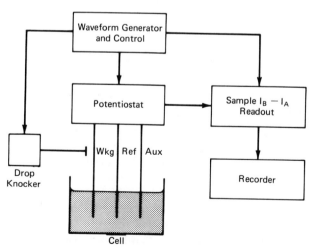

Figure 5-2. Block diagram of a differential pulse polarograph.

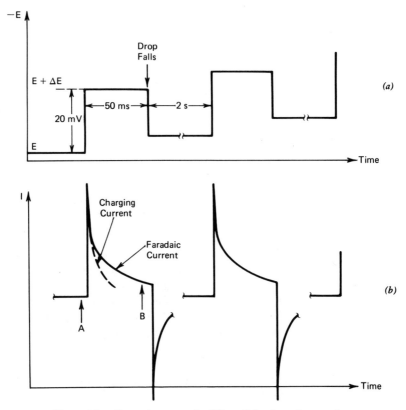

Figure 5-3.   Current response in differential pulse polarography.

as illustrated in Figure 5-4. Note that the peak height of the signal is proportional to the concentration of the analyte, and that the baseline is nearly flat over a large range of concentrations.

CLASSIFICATION OF STEP METHODS

A great amount of ingenuity has been devoted to designing various types of experiments involving voltage steps. Leaving aside certain older techniques that used imperfect sampling, and limiting ourselves to the DME, let us consider the most significant possibilities.

In Figure 5-5 is shown the first fully developed technique, the *square-wave polarography* of Barker [1–3], who used a symmetrical square-wave signal of small amplitude (20 mV) and moderate frequency (225 Hz). The response contains large charging transients that must decay within the rather short periods involved (2 ms). This is a disadvantage, since it requires a low resistance, and hence a rather concentrated supporting electrolyte. On the other hand, the proximity of the sampling

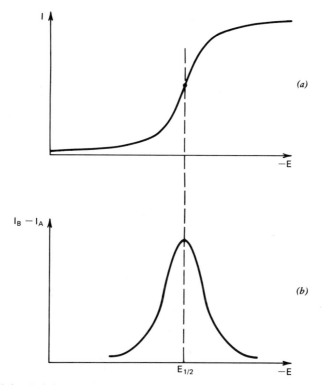

Figure 5-4. Relation between the DC polarographic wave (a) and the DPP peak (b).

points increases the rejection efficiency for interfering signals, and the magnitude of the faradaic current due to the pulse is greater at this short time, so that the sensitivity is outstanding.

In square-wave polarography at this frequency, the faradaic transient from one pulse is still present when the next pulse arrives, so that there is some interference between points. This can be diminished by going to a lower frequency or by leaving a period of relaxation between pulses. The most important technique using this latter approach is *differential pulse polarography*, (DPP) [3-7], discussed briefly above and shown again in Figure 5-6 for comparison. Usually the elapsed time between sampling points A and B is an order of magnitude larger than in square-wave polarography. This makes the elimination of charging-current effects much easier, but as the time base increases, the rejection of noise decreases, so that the two methods have roughly the same sensitivity.

Also useful is the large-step procedure known as *normal pulse polarography* [4, 8-10], illustrated in Figure 5-7. Either single-point sampling (at point B) or two point sampling $(B - A)$ can be employed. In either case, the output gives a step or wave resembling that obtained in DC polarography, with slightly improved performance.

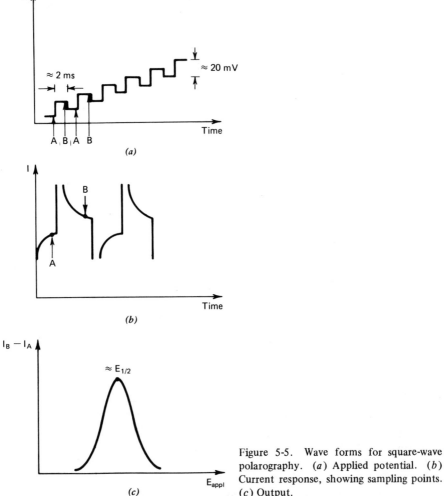

Figure 5-5. Wave forms for square-wave polarography. (*a*) Applied potential. (*b*) Current response, showing sampling points. (*c*) Output.

THE FUNDAMENTAL PROCESS

In the methods discussed above, the fundamental process is a voltage step, with information gathered at the points A and B, before and after the transition. The separation between points is in the range of 2–100 ms, depending on the method.

As with DC polarography, the current contains two major components, faradaic and charging, but now the effect of the voltage step on both must be considered. We shall discuss the various currents separately, assuming no interaction between them. The change in the surface area of the mercury drop between A and B is normally small enough to be neglected, except as a residual effect.

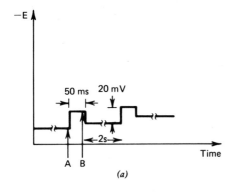

(a)

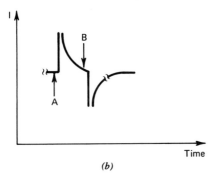

(b)

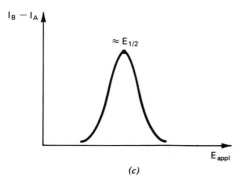

(c)

Figure 5-6. Wave forms for DPP. (*a*) Applied potential. (*b*) Current response, showing sampling points. (*c*) Output.

The currents that must be considered are the following:

1. The faradiac current at *A* (unperturbed).
2. The faradaic current at *B*.
3. The faradaic current transient on going from *A* to *B*.
4. The charging current at *A* (unperturbed).
5. The charging current at *B*.
6. The charging current transient on going from *A* to *B*.

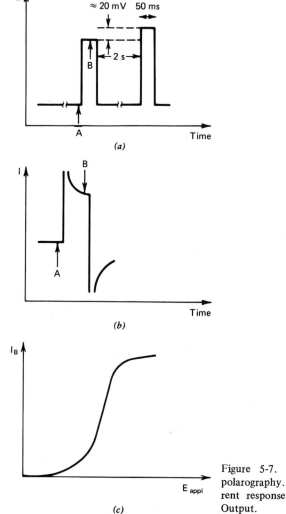

Figure 5-7. Wave forms for normal pulse polarography. (*a*) Applied potential. (*b*) Current response, showing sampling points. (*c*) Output.

The first component to be discussed is the faradaic current, represented in Figure 5-8*b*. It can be thought of as a small segment, length Δt, of an Ilkovič-type curve, $I = (\text{const}) \cdot t^{1/6}$, and can be approximated by a straight line with a small positive slope. The dashed line represents the current corresponding to $E + \Delta E$, which is the Ilkovič curve for the potential at point B. The separation between the two curves is greatest at $E_{1/2}$ and is zero in the diffusion-current region. This means that the effect of previous reductions is mostly eliminated, an important property of differential operation.

It is interesting to estimate the extent of this rejection. The only factor concerned is the effect of the small change of surface area between A and B. If we

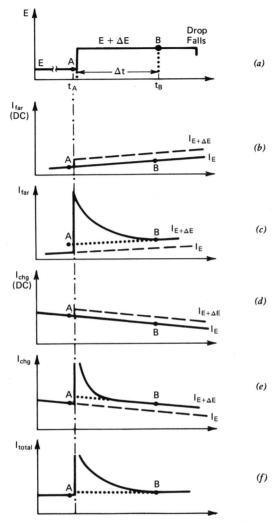

Figure 5-8. Step stimulus (*a*), and responses thereto (*b–f*). The slopes are exaggerated for clarity.

take as typical values the sampling time $t_a = 2$ s, and the interval $\Delta t = 0.05$ s, we can determine the fractional change caused by the voltage step:

$$\frac{I_B - I_A}{I_B} = \frac{(\text{const}) \cdot t_B^{1/6} - (\text{const}) \cdot t_A^{1/6}}{(\text{const}) \cdot t_B^{1/6}}$$

$$= 1 - \left(\frac{2.00}{2.05}\right)^{1/6} = 0.004 \tag{5-1}$$

The effect is small, only 0.4 percent of I_b, but not negligible. Note that as the voltage shifts from E to $E + \Delta E$, the faradaic current does not rise directly to the new value, but follows a transient curve, as shown in Figure 5-8c. The transient decays with a $1/\sqrt{t}$ dependence [11], in accordance with the relation:

$$\Delta I = nFA \left(\frac{D_{OX}}{\pi \Delta t}\right)^{1/2} C_{OX}^* H(E_j, t_j) \tag{5-2}$$

in which a reduction is assumed. The quantity ΔI, the difference between the currents at B and A, is the principal output of the polarograph. The function H depends in a rather complex way on the voltages E_j and time periods t_j of previous steps in the history of the drop. This function is unity if the pulse is large and comes from a region where there is no faradaic process to a point where diffusion currents prevail, as is true in normal pulse polarography. In this instance, the relation reduces to the Cottrell equation. In contrast, the function H plays an important role in square-wave operation.

The effect of the pulse on the charging current must also be considered. Its DC component is shown in Figure 5-8d. This represents the principal factor that limits the sensitivity of DC polarography, since it becomes dominant for solutions of less than about 10^{-5} M. The charging current arises from the growth of the drop, and the need to endow the newly formed surface with its double layer. It depends, therefore, on the applied voltage, and has a different value at E and at $E + \Delta E$. Ths can be seen from the following argument, using the expression for the charge, Q, as a function of the capacitance, C, the area, A, and the voltage, E:

$$Q = CAE \tag{5-3}$$

The charging current is then given by:

$$I_E = \frac{dQ}{dt} = CE \frac{dA}{dt} \tag{5-4}$$

and

$$I_{E+\Delta E} = C(E + \Delta E)\frac{dA}{dt} \tag{5-5}$$

Noting that the area increases as $t^{2/3}$, we obtain:

$$I_{chg} \propto CEt^{-1/3} \tag{5-6}$$

a decreasing function of time. This indicates that longer drop-times are preferable in order to diminish the charging current contribution. In the time interval Δt between A and B, the DC charging current varies little, less than 1 percent in a typical case, including the effect of ΔE. It is therefore expected that DPP should be over

100 times more sensitive than conventional DC polarography. The rejection of DC charging current improves as Δt decreases.

The transition to the new DC charging current takes place through a transient (Figure 5-8e) that is normally quite large but of short duration. Its form follows the behavior of the double-layer capacitance charged through the resistance R of the solution:

$$\Delta I_{chg} = \frac{\Delta E}{R} \exp\left\{ -\frac{1}{CAR} \cdot t \right\} \qquad (5\text{-}7)$$

where C is the capacitance per unit area. The exponential dependence is such that the value decreases rapidly for $t > CAR$. In a typical case, CAR might be about 0.1 ms, so that beyond 1 ms or so the charging transient will have become negligible.

The total current is the sum of those previously discussed and is shown in Figure 5-8f. The dominant contribution to ΔI will be the faradaic transient, with practically no charging current and better than 99 percent rejection of both the diffusion currents from previous reductions and the charging current due to drop growth. A yet smaller contribution derives from the fact that the potential at B is ΔE volts more negative than at A. This is reflected in both DC faradaic and charging components.

These four residual currents represent the limiting factors in both differential

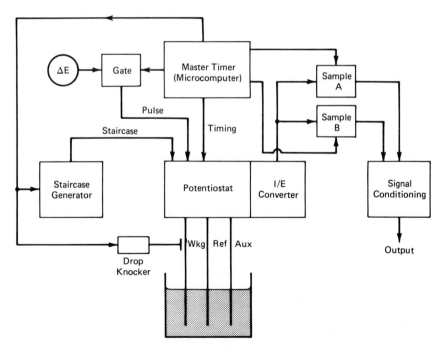

Figure 5-9. An example of pulse instrumentation. The signal-conditioning module furnishes to the recorder the necessary information concerning ΔI and ΔE.

pulse and square-wave polarography. Of the two, square-wave is favored in terms of sensitivity, since it uses shorter times between A and B. In addition, as shown by Rifkin and Evans [11], the useful signal increases for reversible systems if Δt is small. Unfortunately, there has always been much less interest on the part of industry in providing square-wave polarographs than differential pulse units.

INSTRUMENTATION

The instrumentation for step polarography (Figure 5-9) centers around a potentiostat that must have good transient response and a timing system to generate the rather complex sequence of commands needed. This operation can be implemented with a dozen or so integrated circuits perhaps programmed through a microprocessor [12]. One possible sequence of commands [13] is shown in Figure 5-10, where the horizontal lines represent the digital levels ("0" and "1") of the various command signals.

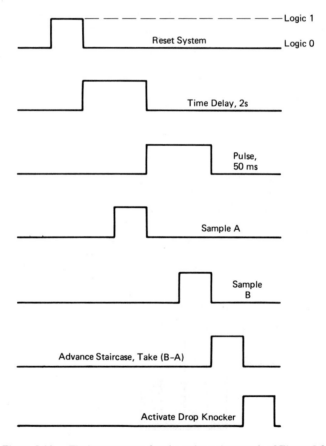

Figure 5-10.   Timing sequence for the pulse polarograph of Figure 5-9.

Once the sequence is established, the instrument must generate the voltage waveform. One popular method is to use a binary counter driving a digital-to-analog converter, which produces a staircase by advancing the voltage by one increment every time a count is received. Synchronized with each step is a command that allows the voltage pulse, ΔE, to be applied at the proper time and other commands to ensure that the current is read at the correct points, A and B. These signals are implemented through analog transmission gates.

The current values at times A and B are fed into two sample-and-hold amplifiers that store the corresponding values, passing them on in turn to a subtraction element to give the desired output, ΔI, to a recorder.

REFERENCES

1. P. E. Sturrock and R. J. Carter, *CRC Crit. Rev. Anal. Chem.*, **1975**, *6*, 201.
2. G. C. Barker and I. L. Jenkins, *Analyst (London)*, **1952**, *77*, 685.
3. G. C. Barker, *Anal. Chim. Acta.*, **1958**, *18*, 118.
4. A. A. A. M. Brinkman and J. M. Los, *J. Electroanal. Chem.*, **1964**, *7*, 171.
5. E. P. Parry and R. A. Osteryoung, *Anal. Chem.*, **1965**, *37*, 1634.
6. H. E. Keller and R. A. Osteryoung, *Anal. Chem.*, **1971**, *43*, 342.
7. I. Ružić, *J. Electroanal. Chem.*, **1977**, *75*, 25.
8. E. P. Parry and R. A. Osteryoung, *Anal. Chem.*, **1964**, *36*, 1366.
9. K. B. Oldham and E. P. Parry, *Anal. Chem.*, **1966**, *38*, 867.
10. D. J. Myers, R. A. Osteryoung and J. Osteryoung, *Anal. Chem.*, **1974**, *46*, 2089.
11. S. C. Rifkin and D. Evans, *Anal. Chem.*, **1976**, *48*, 1616.
12. E. B. Buchanan, Jr., and W. J. Sheleski, *Talanta*, **1980**, *27*, 955.
13. B. H. Vassos and R. A. Osteryoung, *Chem. Instrum.*, **1974**, *5*, 257.

Chapter 6

VOLTAMMETRY:
III. AC POLAROGRAPHY

It was remarked previously that voltammetry can be considered to be largely the study of the equivalent circuit of the electrode-solution interface. Since this interface has significant capacitive properties, more information can be obtained from the electrode *impedance* than from its resistance alone. To make such a study, we can apply a small alternating signal to the cell along with the usual DC ramp or staircase and examine its effects on the cell current. There are a number of closely related techniques that make use of this approach; they are special cases of the field of AC polarography.

The signal-to-noise (S/N) relations are favorable because the information about the chemical system is carried in a narrow band of frequencies only, and extraneous signals at other frequencies can be effectively eliminated. This is an example of AC modulation, much used in various fields to increase the S/N ratio.

We shall see that AC polarography permits an increase in sensitivity compared to the conventional DC method, so that the detection limits in favorable cases can be reduced to about $10^{-8}\,M$. An interesting feature is that the sensitivity decreases drastically as the redox system becomes less reversible. This may be advantageous if a reversible system is partially masked by one that is irreversible.

THE BASIC EXPERIMENT

The electrodes and cell configuration of DC polarography are usable in the AC mode without modification. Let us apply an excitation to the cell combining the DC staircase with a small AC voltage (say 10 mV at 150 Hz). The electronic system will extract the AC component of the cell current and display this response as a function of the DC voltage.

One effect of the AC modulation is to convert the polarogram from the step-

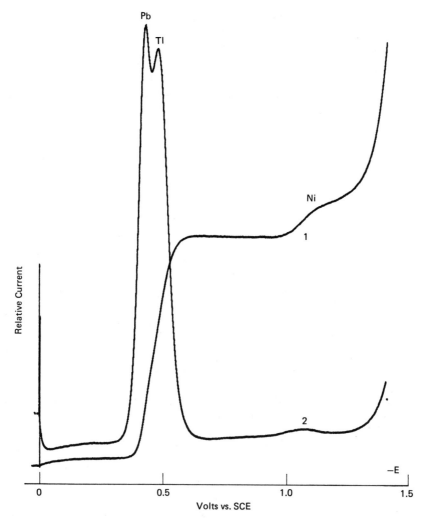

Figure 6-1. Polarograms of 0.5 M KCl containing 0.0001 M Pb^{++} and Ni^{++} and 0.0002 M Tl^{+}. Curve 1: DC polarogram. Curve 2: AC polarogram. Note that the lead and thallium are not resolved in the DC mode but are easily visible in AC, and that nickel shows much reduced sensitivity in AC, indicating partially irreversible behavior. (Recorded on a Sargent-Welch Polarograph.)

like waves, seen in the DC mode, to a derivative curve, similar to those encountered in DPP. Figure 6-1 shows the output signal of such an experiment, compared with a conventional polarogram. The maximum of each peak is called a *summit*, and the corresponding coordinates are the *summit potential*, E_s, and the *summit current*, I_s. The summit potential coincides with $E_{1/2}$, and measurements at various concentrations show that I_s varies linearly with concentration. Another characteristic is the virtual suppression of peaks corresponding to irreversible reactions.

THEORETICAL CONSIDERATIONS

Let us assume that our cell contains KCl as supporting electrolyte, together with a small amount of Pb^{++}, forming a nearly reversible system. As the DC voltage increases, the relative concentrations of OX and RED at the electrode interface will vary, as in DC polarography, from all OX, no RED, to the reverse when the diffusion plateau is reached. However, this is now merely a first approximation because of the presence of the AC modulation.

If we look closer, we will realize that the applied potential fluctuates around its nominal value at any point along the ramp, at the AC frequency. The applicable form of the Nernst equation:

$$E = E_{1/2} + \frac{0.0592}{n} \log (C_{OX}^s / C_{RED}^s) \tag{6-1}$$

shows that as the applied voltage E varies, so must the OX/RED ratio. One can perhaps visualize some of the lead atoms swinging back and forth from one oxidation state to the other 150 times a second.

This means that the faradaic current, which alone is responsible for redox processes, must also fluctuate at the 150-Hz frequency, provided that both states of lead are present. In the regions where $E \gg E_{1/2}$ or where $E \ll E_{1/2}$, only one form is present, Pb_{aq}^{++} on the positive side, $Pb_{(Hg)}^0$ on the negative, so that no AC faradaic current can pass. This accounts for the flat background on either side of the summit.

There is, of course, a nonfaradaic AC current flowing at all times as a result of the double-layer capacitance, that is not affected by the faradaic process. The net alternating current is the sum of the faradaic and nonfaradaic components, and this will go through a maximum at the half-wave potential, where the concentrations of OX and RED are equal. The resulting AC polarogram is that already seen in Figure 6-1. Just as in DC, a sampling circuit is usually employed to give a step-wise approximation to the envelope.

If, rather than lead, we select nickel to study, the results will be quite different. Since the system is somewhat irreversible, a considerable negative overpotential must be applied to force the reduction to take place, and a similar overpotential of opposite sign is necessary to reoxidize the neutral atoms. This is tantamount to saying that the exchange currents are small (Figure 6-2), whereas for the reversible system they are much greater. Another way of looking at this is to say that in a of the figure, the faradaic impedance, Z_{far}, is high, while in b it is low. Hence, very little alternating current, $I_{ac} = E_{ac}/Z_{far}$, flows in the irreversible case, contrasted with the much greater current if the system is reversible. Experimentally, the response is reduced to a small bump, as seen for nickel in Figure 6-1, or may disappear entirely with a completely irreversible reaction. This effect provides one method of establishing whether or not a system is reversible.

To understand the phenomena taking place in AC polarography, it is essential

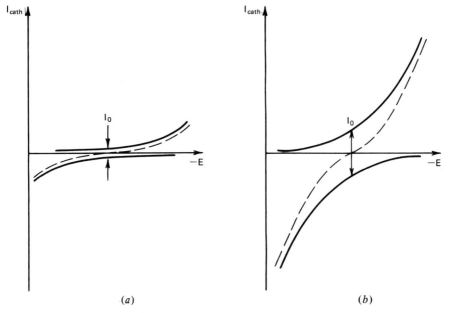

Figure 6-2. Exchange currents for (a) an irreversible, and (b) a reversible system. The slope of the total current curve (dashed lines) is the faradaic admittance, the reciprocal of the faradaic impedance.

to grasp the impedance and phase relations at the interface, and this is most readily done by means of an equivalent electrical circuit. Since the impedance of the double-layer capacitance is relatively small at the frequency used, appreciable AC will pass through the interface via this route, without producing any faradaic effect. This capacitance can therefore be considered to be in parallel with the faradaic impedance, and both to be in series with the solution resistance R, as in Figure 6-3 [1, 2].

AC theory tells us that when an alternating voltage is impressed upon a capacitor, the resulting current *leads* the voltage by a phase angle of $90°$ (Figure 6-4), whereas for a pure resistance the current and voltage are in phase, $(0°)$. In contrast, the faradaic impedance can be shown to produce an intermediate phase shift of $45°$ (for reversible systems) [3]. This circumstance provides a means for distinguish-

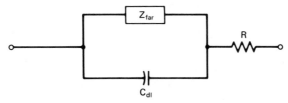

Figure 6-3. Equivalent circuit for the working electrode.

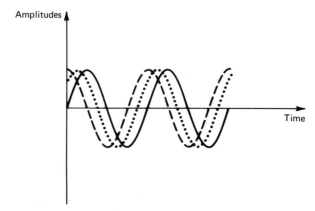

Figure 6-4.   Phase relations in AC polarography. Solid line: applied sine-wave voltage; dashed line: charging current (a cosine wave); dotted line: the faradaic current. The amplitudes are shown as equal for ease of comparison.

ing between faradaic and nonfaradaic currents flowing through the equivalent circuit of Figure 6-4.†

These phase relations are depicted vectorially in Figure 6-5a [3]. The arrow pointing upward represents the applied AC potential and is taken as the reference vector. In the absence of resistance, the charging current passing through the double-layer capacitance is represented by a horizontal vector, and the faradaic current by one at 45°. Hence, an instrument so designed that it responds only to current that is in phase with the applied voltage will, in principle, eliminate the charging current and still give an output that represents 70 percent ($\sqrt{2}/2$) of the maximum faradaic current.

In the presence of significant resistance, the phase shifts due to both faradaic and capacitive currents are altered by approximately the same amount, Ψ, as shown in Figure 6-5b. The discrepancy angle, Ψ, can be allowed for by adjustment of a phase-shift control. This adjustment is accomplished by setting the DC voltage at a point where no polarographic wave is present and adjusting for minimum (ideally zero) pen response. Even so, the discrimination against charging current is never perfect, due largely to the fact that most redox systems deviate to some extent from reversibility.

As a result of the AC faradaic current, the concentration profiles also assume sinusoidal forms, characterized by their own phase angles, θ.

The mathematical expressions for the summit current, I_s, and the phase angle, ϕ, between the applied voltage and the faradaic current, can be derived from the Fick diffusion equations by application of the particular boundary conditions:

---

†Since Z_{far} gives a 45° shift and C_{dl} a 90° shift, in principle these are separable, but only if the series resistance R is negligible. This analogy of a network of resistances and capacitances should not be taken to be an exact representation of the actual cell. The particular values of the components are valid only for a specific frequency and DC voltage.

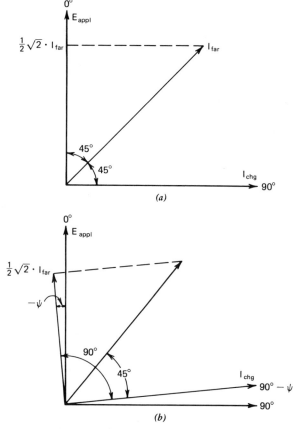

Figure 6-5. Phase relations of charging current (I_{chg}) and faradaic current (I_{far}) to the applied alternating voltage in AC polarography. (*a*) the approximate relations if the series resistance is negligible. (*b*) The corresponding diagram for a real cell; the angle Ψ must be corrected for by the phase adjustment control.

$$C_{OX}^{s} = \Delta C_{OX}^{s} \sin (\omega t + \theta_{OX})$$

$$C_{RED}^{s} = \Delta C_{RED}^{s} \sin (\omega t + \theta_{RED}) \tag{6-2}$$

where ΔC_{OX}^{s} represents the amplitude of the sinusoidal fluctuations in the concentration of OX, the quantity ΔC_{RED}^{s} is similarly defined, θ is the phase angle between the concentration fluctuation and the applied potential, and ω is the angular frequency. The derivation is too long and involved to present here, but it can be found in [1, 4]. The final results (for the low frequencies that are useful in polarography) are given by:

$$I_{AC} = \frac{n^2 F^2 A \Delta E C_{OX}^{s} (\omega D_{RED})^{1/2} \exp \{nFE/RT\}}{RT[(D_{RED}/D_{OX})^{1/2} + \exp \{nFE/RT\}]^2} \tag{6-3}$$

and

$$\phi = \cot^{-1} \left(1 + \frac{(2\omega D_{RED})^{1/2} \exp\{\alpha nFE/RT\}}{k[(D_{RED}/D_{OX})^{1/2} + \exp\{nFE/RT\}]} \right) \tag{6-4}$$

where ΔE is the amplitude of the applied AC voltage, and k is the applicable heterogeneous rate constant.

Certain conclusions can be drawn from these equations. Note that the (faradaic) current in Eq. (6-3) is directly proportional to ΔE as well as to C_{OX}. Also, the current increases with the square root of the frequency (whereas the charging current increases as the first power of the frequency). This would imply that the sensitivity of the method is greater the larger the amplitude and frequency of the AC modulation. However, the derivation of the equations was predicated on both being small enough so that certain approximations could be made. In practice, ΔE is usually between 1 and 30 mV, and the frequency below about 300 Hz (the power line frequency can be used satisfactorily).

In Eq (6-4), k is normally greater than $(2\omega D_{red})^{1/2}$, so that the second term approaches zero and $\phi \longrightarrow \cot^{-1}(1) = 45°$.

INSTRUMENTATION

The basic structure of an AC polarograph is shown in Figure 6-6. The AC from the oscillator is combined with the DC ramp or staircase and impressed on the potentiostat, amplifier 1. The cell current, containing both AC and DC components, is converted by amplifier 2 to a corresponding voltage. The combination of R and C acts as a high-pass filter, permitting the AC portion of the signal to pass, but suppressing the DC component. The AC signal is then rectified to give a proportional DC, and recorded.

In order to take advantage of the phase relations demonstrated in Figure 6-5, a phase-sensitive detector must be included, as shown in Figure 6-7. This is a device that compares the AC signal with a reference at the same frequency obtained directly from the oscillator, and acts as a gate to transmit to the recorder only that

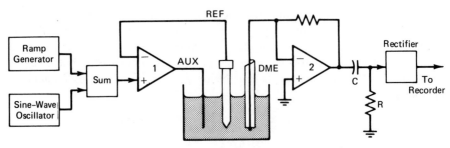

Figure 6-6.   A basic AC polarograph.

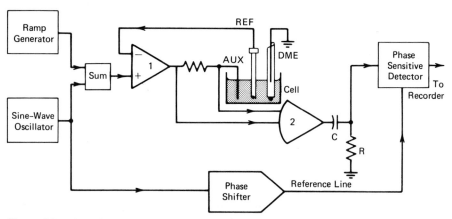

Figure 6-7. An AC polarograph with phase-sensitive detector. Amplifier #2 is shown as an instrumentation or differential amplifier, measuring the potential drop across the resistor in the auxiliary electrode line.

component of the signal that bears a specific phase relation. In order to compensate for the phase changes introduced into the signal by various electronic components and by the cell itself, an adjustable phase-shifting device must be included in the reference connection.

The phase-sensitive detector also acts as a filter, discriminating against frequencies other than that of the oscillator, including harmonics. Since electrical noise occurs at all frequencies, most of it is eliminated by this filtering action. In this context, the double-layer charging current can be considered to be noise.

Second-Harmonic AC Polarography

When a pure sine-wave voltage is impressed upon an electrical circuit component, the current that passes may or may not also be a sine wave. If the component is linear, then the response is still a sine wave, whereas, for nonlinear components, it is subject to *harmonic distortion*, which consists in the generation of multiples of the original frequency. Figure 6-8 shows how distortion can originate in the polarographic cell.

The faradaic impedance at the DME constitutes a nonlinear circuit element so that the alternating portion of the faradaic current contains an appreciable percentage at the second harmonic frequency—twice the frequency of the applied voltage. (Contributions of higher harmonics can be neglected.) On the other hand, the double-layer capacitance is a nearly linear element, and thus does not contribute appreciably to the harmonic content. This means that a polarograph that responds to the second harmonic, rather than to the entire AC current, will discriminate strongly against the charging current.

The nonlinearity that gives rise to harmonic distortion appears both prior to and following the summit potential, but nearly disappears right at the summit and also on the horizontal segments of the curve. Consequently, the second-harmonic out-

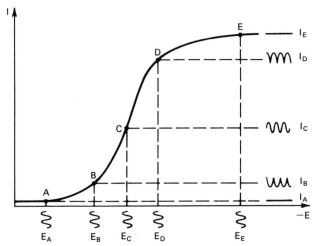

Figure 6-8.   The source of harmonic distortion in AC polarography. Small sinusoidal voltages are shown at five locations. At E_C (same as $E_{1/2}$), the polarogram is nearly linear, and no distortion results. At E_B and E_D, where the curvature is marked, the current shows considerable distortion. At E_A and E_E, the curve is so nearly horizontal that no sinusoidal current variation is seen at all.

puts exhibits a zero at $E_{1/2}$ flanked by two maxima, S_1 and S_2. This is shown in Figure 6-9, together with other cases, for a reversible system.

Since phase-sensitive detection is so useful in fundamental AC polarography, it is of interest to investigate its applicability to the second-harmonic case. We cannot here take the applied alternating voltage as a reference vector because phase differences can only be defined as relating variables of the *same* frequency. Hence, the best we can do is to look at the *relative* phases of the second-harmonic current in the two branches of the curve of Figure 6-9c.

In order to instrument second-harmonic polarography, the basic circuit of Figure 6-7 can be used, with the sole addition of a component to go between the oscillator and the phase shifter, to double the frequency in the reference line. This is shown in Figure 6-10. The result of an experiment with this modified instrument gives a curve such as that in Figure 6-9d. This shows one of the branches as inverted, indicating a phase difference of 180° between the branches, in agreement with theoretical predictions [4-6]. An example is presented in Figure 6-11.

The mathematical theory of second-harmonic AC polarography [4, 6] is even more complex than that of the fundamental case. It leads to equations that are quite intractable unless solved by computer. The final equation [6] shows that the current is in direct proportion to the bulk concentration of the reducible species and to the square of the applied AC voltage amplitude. It also increases as the frequency is increased, but by a nonlinear function. Theory correctly predicts that, for a reversible process, the phase-detected current will be of one sign for potentials more positive than the half-wave potential, and the opposite sign for more negative potentials, passing through zero at $E_{1/2}$. The two summits are separated by $68.4/n$ mV [4].

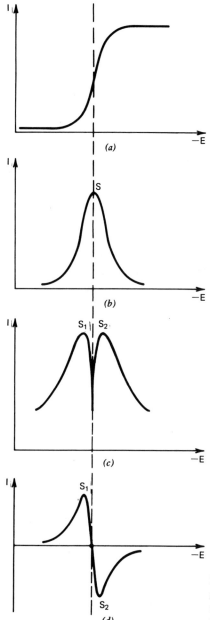

Figure 6-9. Comparison of sampled polarograms; (a) DC, (b) simple AC, (c) second-harmonic AC, (d) second-harmonic AC with phase-sensitive detection.

The most convenient measure of concentration is the vertical distance between the two summits, since this automatically corrects for any error in placement of the background or residual current level.

The dependence of the output on the square of the applied AC amplitude suggests the benefit of using as large a voltage as possible, but in practice it must

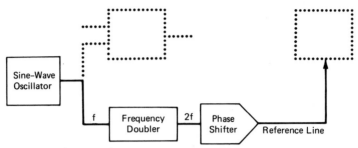

Figure 6-10.   Modification to the circuit of Figure 6-7 to allow for phase-sensitive detection at the second-harmonic frequency.

be limited to about 10 mV in order to avoid degradation of resolution. The frequency can be varied over a wider range than in the fundamental AC case. For example, the reduction of Cd^{++} to Cd_{Hg} in a neutral medium shows an increase in summit current from about 0.9 μA at 23 Hz to 3.2 μA at 2220 Hz [7]. On the other hand, the same system shows no change in the voltage separation of the two summits [8]. It must be emphasized that the preceding discussion applies only to systems that show reversible characteristics when measured at DC. Others will

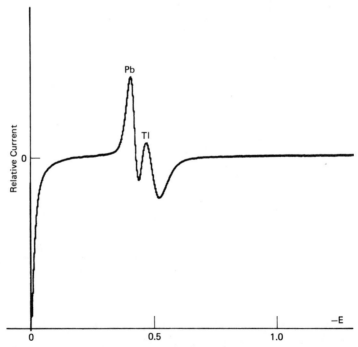

Figure 6-11.   Second-harmonic, phase-detected AC polarogram. The solution is the same used for Figure 6-1. Note that the nickel does not show up at all. (Recorded on a Sargent-Welch Polarograph.)

give complicated deviations from the basic equations [2, 6], and the suitability of the method for specific cases must be determined experimentally.

Methods Related to AC Polarography

As we have seen, an alternating voltage impressed on a nonlinear circuit element gives rise to harmonics of the applied frequency. The current can be expressed by a Taylor's series.† If the excitation is given by $E_{AC} = E_0 + A \cos \omega t$, then the AC response is $I_{AC} = E_{AC}/Z_{far}$. Expansion then gives:

$$I_{AC} = b_1 \cos (\omega t + 45°) + b_2 \cos^2 (\omega t + 45°)$$
$$+ b_3 \cos^3 (\omega t + 45°) + \cdots \quad (6\text{-}5)$$

where the bs include all constant coefficients. The angle $45°$ is the phase shift due to the faradaic impedance. This series rapidly converges, which means that $b_1 > b_2 > b_3 > \cdots$, and for the present case, the third and higher harmonic terms can be neglected. The second term can be restated in a more useful form by making use of the trigonometric identity

$$\cos^2 x = \tfrac{1}{2} + \tfrac{1}{2} \cos 2x \quad (6\text{-}6)$$

so we can write:

$$I_{AC} = b_1 \cos (\omega t + 45°) + \tfrac{1}{2} b_2 + \tfrac{1}{2} b_2 \cos (2\omega t + 90°) \quad (6\text{-}7)$$

The first term refers to the fundamental AC polarographic signal. The last term, in $\cos 2\omega t$, describes the second harmonic signal. This leaves the invariant term $\tfrac{1}{2} b_2$, which must correspond to a DC component of the current.

This can be visualized by reference to Figure 6-12. Since this DC component and the term in $2\omega t$ arise from the cosine-squared term of the expansion, they should be expected to carry the same information about the electrochemical system. This is a form of rectification, and its measurement is called *faradaic rectification polarography*.

Another method that utilizes the nonlinear nature of the electrode process is known as *intermodulation polarography*. This requires the simultaneous application of two different frequencies, f_1 and f_2. The action of the non linear circuit is to generate the sum and difference frequencies, $f_1 + f_2$ and $f_1 - f_2$. For example, if two oscillators connected to the input of the potentiostat are set at 23 Hz and 300 Hz, respectively, then the cell current will contain components at 277 and 323 Hz in addition to the unchanged 23- and 300-Hz frequencies. Measurement with the aid of a suitably tuned amplifier and detector provides information similar to that of second-harmonic AC polarography.

For analytical purposes, these methods have little advantage [9].

---

† This approach can be shown to be equivalent to the use of the more elaborate Fourier series.

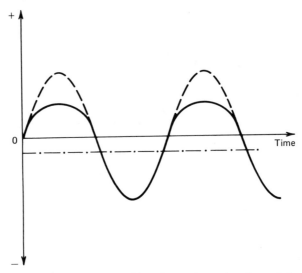

Figure 6-12.   A distorted sine wave resulting from a decreasing slope of the current–voltage curve.   The dashed curve is the original sine wave, the solid curve shows the distortion effects. The dash–dot line represents the DC current due to faradaic rectification.   Additional components of higher frequencies are also generated.

AC-Pulse Polarography

It is possible to combine the principles of AC polarography with those of pulse techniques in such a way as to retain the beneficial features of both. This has been reported for the "normal" pulse case [10], but to date has been found most useful with the differential pulse method [11, 12].

The potential applied to the electrode as a function of time is shown in Figure 6-13 [11]. This can be regarded as AC polarography wherein the DC ramp is replaced by the pulsed ramp as used in DPP. The requirements are simply that there be many cycles of AC modulation during each pulse, and that the AC amplitude, ΔE, be less than the pulse height, E_p.

The measurement procedure, just as in DPP, is to sample the current (AC in this case) just *before* the pulse (I_a) and again near the end of the pulse (I_b). The theory [11] then shows that the difference $I_b - I_a$ is proportional to the bulk concentration of the active species, to the amplitude ΔE, and to the square root of the frequency. The shape of the polarogram (with phase-sensitive detection) resembles the phase-detected second-harmonic polarogram.

The charging current produced by the pulse almost completely dies out before the current is sampled, as in DPP. The effect of the small residuum is merely to offset the background level slightly, and even this effect is eliminated by making quantitative measurements vertically between the two summits.

This makes another sensitive and useful method for the determination of low concentrations of species with reversible or quasi-reversible electrode processes.

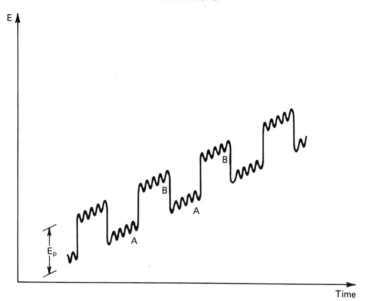

Figure 6-13. Representation of the potential excitation function in differential pulse AC polarography. Measurements are taken at points *A* and *B*. [Redrawn from Analytical Chemistry (11).]

The limit of detection for a process such as the reduction of Cd^{++} in KNO_3 is reported [11] to be about 5×10^{-8} *M*. It is not applicable to irreversible systems.

REFERENCES

1. B. Breyer and H. H. Bauer, "Alternating Current Polarography and Tensammetry," Wiley-Interscience, New York, **1963**, p. 56ff.

2. D. E. Smith, *CRC Crit. Rev. Anal. Chem.*, **1971**, *2*, 247.

3. A. M. Bond, *Anal. Chem.*, **1972**, *44*, 315.

4. D. E. Smith, "AC Polarography and Related Techniques," in "Electroanalytical Chemistry" (A. J. Bard, Ed.), Dekker, N.Y., **1966**, Vol. 1.

5. D. E. Smith, *Anal. Chem.*, **1963**, *35*, 1811.

6. T. G. McCord and D. E. Smith, *Anal. Chem.*, **1968**, *40*, 289.

7. T. G. McCord and D. E. Smith, *Anal. Chem.*, **1969**, *41*, 131.

8. H. Blutstein, A. M. Bond, and A. Norris, *Anal. Chem.*, **1974**, *46*, 1754.

9. H. Blutstein, A. M. Bond, and A. Norris, *Anal. Chem.*, **1976**, *48*, 1975.

10. D. E. Smith, A. M. Bond and B. S. Grabaric, *J. Electroanal. Chem.*, **1979**, *95*, 237.

11. A. M. Bond and R. J. O'Halloran, *Anal. Chem.*, **1975**, *47*, 1906.

12. A. M. Bond, B. S. Grabaric, R. D. Jones and N. W. Rumble, *J. Electroanal. Chem.*, **1979**, *100*, 625.

Chapter 7

VOLTAMMETRY: IV. LINEAR SWEEP

It is convenient to distinguish two types of voltammetric experiments which are respectively, dependent on or independent of, the scan rate. The various types of polarography belong to the second category, since the results of the measurement do not depend on how fast the voltage range is scanned. One could even jump at random between voltages without changing the current–voltage relationships, as long as the voltage is constant during the life of each mercury drop.

In contrast, in the method to be described in this chapter, the measurements *do depend* on the rate at which the voltage is scanned. The current observed also depends on the exact form of the scan, which is usually a simple linear sweep.

Rate-dependent voltammograms are obtained with solid electrodes or stationary mercury drops, rather than with the DME. In spite of this, the term *stationary-electrode polarography* has sometimes been used to describe this method.† Some authors argue that the method is, strictly speaking, neither polarography nor voltammetry, and that it properly should be called *chronoamperometry with linear potential sweep*, a wonderfully descriptive but too lengthly title. We shall use the name *linear sweep voltammetry (LSV)*.‡ The term *cyclic voltammetry (CV)* describes a variant in which the potential is scanned back and forth between two limits. The method in its modern form owes much to the efforts of Shain and Nicholson [3-5].

THE BASIC EXPERIMENT

Consider the instrument of Figure 7-1, in which a signal generator produces a voltage sweep from E_i to E_f. A potentiostat applies it to the electrode, which can be a

---

†A method devised by Heyrovský in 1941 used nonlinear scans. This was called *oscillographic polarography*, a name later applied by Randles and Ševčík [1, 2] to describe a linear sweep operation using an oscilloscope.

‡The use of this term does not imply that other forms of voltammetry may not employ a linear scan.

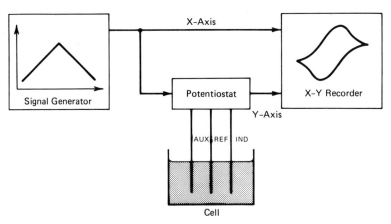

Figure 7-1.   Block diagram of an instrument for cyclic voltammetry.

solid, such as carbon or platinum, or a stationary mercury drop. The SMDE electrode, described in Chapter 4, can be used conveniently.

An example of the resulting current for a reversible system is shown in Figure
7-2. A characteristic peaked curve is obtained, occuring in the region of the polarographic wave. Note that $E_{1/2}$, as defined in polarography, falls about midway between the half-peak potential $E_{pk/2}$, and the peak itself, E_{pk}. The value $E_{1/2}$ can
be estimated from a linear sweep polarogram, even though it is not directly related
to it.

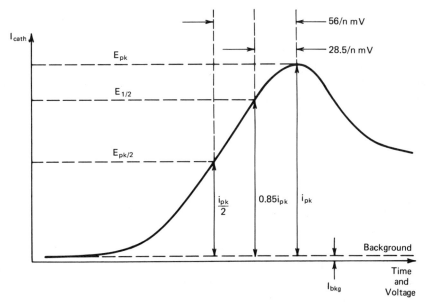

Figure 7-2.   A typical linear-sweep voltammogram for a reversible system. The potential differences indicated are valid at 25°C. The symbol i_{bkg} denotes the residual current background
at E_{pk}. The quantity $E_{pk/2}$ is defined as the potential where $I = \frac{1}{2} I_{pk}$.

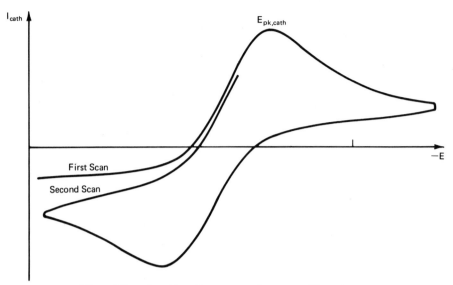

Figure 7-3.   A cyclic voltammogram for a reversible system.

The descending part of the curve follows a $1/\sqrt{t}$ dependence, so that for fast scans the current has a shorter time in which to decrease, and the curve decays less than in a slower scan. The peak height changes proportionally to the square root of the scan rate, v, as well as to the concentration of the active species.

For cyclic voltammetry, the scan is continued in the reverse direction from E_f back to E_i, and a downward-going curve results. The usual practice is to represent it inverted on the same graph, as illustrated in Figure 7-3, by plotting against voltage rather than time.

The two peaks in Figure 7-3, describing the reduction (upper peak) and oxidation (lower peak), are separated by about $60/n$ mV. The two small vertical transitions occurring at E_i and E_f represent the contribution of the charging current:

$$I_{chg} = CA\frac{dE}{dt} = CAv \qquad (7\text{-}1)$$

where C is the electrode capacitance per unit area, and A is the area of the electrode. At the end of the scan, the value of v changes suddenly to $-v$, and a current jump of $2I_{chg}$ is observed. The effect is more pronounced at high scan rates. Subsequent cycles fail to give identical currents because of changes in concentrations at the electrode surface.

The presence of a second stage of reduction generates a second pair of peaks, as shown in Figure 7-4. The dashed line represents the curve that would have been obtained in the absence of the second peak and constitutes its baseline. Unfortunately compensation for the base is not straightforward in this case, but modern computer techniques permit such deconvolution, as reported by Gutknecht and Perone [6].

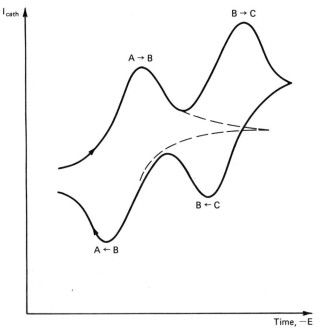

Figure 7-4.   A cyclic voltammogram for a two-step redox system of the type A⟶B⟶C, for example, Cu(II)⟶Cu(I)⟶Cu(0).

THEORETICAL CONSIDERATIONS

In this section we will attempt to show the relation between a voltage sweep stimulus and the electrode response. The interplay between diffusion and electron-transfer kinetics will be examined in some detail. This parallels the treatment in Chapter 4, but differs from it, since in polarography the voltage is constant during each drop from the DME.

The current in a voltammetric experiment is, in general, the resultant of several components:

$$I_{tot} = I_{far} + I_{chg} + I_{ads} + I_{cross} \qquad (7\text{-}2)$$

which describe the contributions of the faradaic, charging, and adsorption currents, as well as a cross term arising from the interaction between the first three. Only the faradaic current is of analytical interest, and the other terms can usually be considered as noise or background. The resolution, sensitivity, and general usefulness of voltammetry depend, in large measure, on the possibility of minimizing or correcting for such background contributions.

The charging current in LSV is given by:

$$I_{chg} = \frac{dQ}{dt} = \frac{d}{dt}(CAE) = CA\frac{\partial E}{\partial t} + AE\frac{\partial C}{\partial t} \qquad (7\text{-}3)$$

where Q is the charge on the electrode, A the area (which is constant), C the capacitance per unit area, and E the voltage relative to the PZC. The term in $\partial C/\partial t$ arises from the fact that, as the scan proceeds, the capacitance of the electrode changes. Both terms depend on the scan rate $v = dE/dt$.

An adsorption contribution is present if the amount of charged species adsorbed varies during the scan. This effect can be written as:

$$I_{ads} = nFA \frac{d\Gamma}{dt} \tag{7-4}$$

where the symbol Γ denotes the surface excess of the particular charged species involved. At higher scan rates, $d\Gamma/dt$ is larger. The current may be of either sign, depending on whether cations or anions are adsorbed or desorbed.

The background current can be subtracted using a separate blank run, but the correction is valid only if $I_{cross} = 0$. The presence of a cross term [7] signifies that the faradaic current has some effect upon the background current, so that the correction procedure is vitiated. Hence a baseline derived by extrapolation from the curve itself may be preferable.

The faradaic current is the result of the requirement that the ratio of OX/RED be altered when the potential is shifted. The Nernst equation, for a reversible system, can be solved for the ratio of surface concentrations:

$$\frac{C_{OX}^s}{C_{RED}^s} = \exp\left\{ \frac{nF}{RT}(E - E^{\circ\prime}) \right\} \tag{7-5}$$

Irreversible processes do not obey this equation.

In an LSV experiment, the potentiostat sweeps the applied voltage from an initial value E_i, at rate v, so that $E = E_i + vt$, and the concentration ratio changes with time:

$$\begin{aligned}
\frac{C_{OX}^s}{C_{RED}^s} &= \exp\left\{ \frac{nF}{RT}(E_i - E^{\circ\prime} - vt) \right\} \\
&= \left(\exp\left\{ \frac{nF}{RT}(E_i - E^{\circ\prime}) \right\} \cdot \exp\left\{ -\frac{nF}{RT}vt \right\} \right)
\end{aligned} \tag{7-6}$$

or

$$\frac{C_{OX}^s}{C_{RED}^s} = g \cdot e^{-mt} \tag{7-7}$$

where g and m are constants defined by reference to Eq. (7-6). This is shown in Figure 7-5a.

Consequently, as the voltage is swept, OX will be converted into RED, as can be

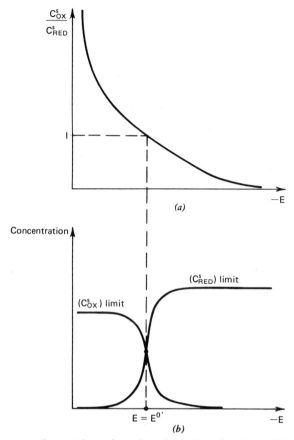

Figure 7-5. Concentrations at the surface of an electrode as functions of time. (*a*) Ratio of concentrations (compare Table 4-1); (*b*) Variation of the individual concentrations as the scan proceeds.

seen in Figure 7-5*b*. The two limiting concentrations are shown unequal, contrary to what one might expect in terms of conservation of mass. In the example shown, the difference is caused by the rapid diffusion of OX toward the electrode. With the transformation into the reduced form, the diffusion coefficient becomes smaller, so that RED does not move away as fast. The electrode process eventually reaches a dynamic equilibrium:

$$\frac{C_{OX}^{s} \text{ (anodic limit)}}{C_{RED}^{s} \text{ (cathodic limit)}} = \left(\frac{D_{RED}}{D_{OX}}\right)^{1/2} \tag{7-8}$$

This effect is often negligible, but it is important if the diffusion coefficients are widely different. One of the consequences is that, if for example we start with a solution of 0.01 *M* OX and run a cyclic voltammogram, there is no guarantee that at the start of the reverse scan the region around the electrode will be 0.01 *M* in

RED. The square-root term in Eq. (7-8) parallels a similar correction term in polarography, but here the effect is more important.

The changes produced at the electrode by the potential sweep extend into the solution to a depth δ, the *diffusion layer*, that is commensurate with $2\sqrt{Dt}$. Beyond about $4\sqrt{Dt}$, the solution remains essentially unchanged. For example, in a rather lengthy scan lasting 1 min, the thickness is less than 1 mm, which can be considered an upper limit. The electrode senses only the concentration C^s at its surface, while the current is determined by the concentration gradient, $(dC/dx)^s$, as defined by Fick's first law:

$$I = nFA\,\Phi^s = nFAD_{OX}\left(\frac{\partial C}{\partial x}\right)^2 \tag{7-9}$$

where Φ^s is the flux of active species at the electrode surface.†

In general, if there is no adsorption involved, the fluxes of RED and OX are much larger than the rate at which they accumulate at the electrode surface. Thus, one can equate the two fluxes:

$$\Phi^s_{OX} = D_{OX}\left(\frac{\partial C_{OX}}{\partial x}\right)^s = -\Phi^s_{RED} = -D_{RED}\left(\frac{\partial C_{RED}}{\partial x}\right)^s \tag{7-10}$$

This equation and Eq. (7-7) are the two fundamental relations that determine the response of a reversible electrode in LSV.

If the electron transfer is slow (an irreversible process), the flux is controlled by the electron-transfer rate (as seen in Chapter 2):

$$\Phi^s = r_t = k^\circ C^s_{OX}\exp\left\{-\frac{\alpha nF}{RT}(E - E^{\circ\prime})\right\}$$
$$- k^\circ C^s_{RED}\exp\left\{\frac{(1-\alpha)nF}{RT}(E - E^{\circ\prime})\right\} \tag{7-11}$$

which introduces two additional parameters, α and k°. Eqs. (7-10) and (7-11) can be combined to give the dual equality that determines an irreversible process:

$$r_t = D_{OX}\left(\frac{\partial C_{OX}}{\partial x}\right)^s = -D_{RED}\left(\frac{\partial C_{RED}}{\partial x}\right)^s \tag{7-12}$$

Another case to be considered is that of adsorption. The pertinent equation is:

---

†One can think of a voltammogram as a plot of $(\partial C/\partial x)^s$ vs. time, since this is the *only* solution property that is actually measured.

$$D_{OX}\left(\frac{\partial C_{OX}}{\partial x}\right)^s - \frac{\partial \Gamma_{OX}}{\partial t} = -D_{RED}\left(\frac{\partial C_{RED}}{\partial x}\right)^s + \frac{\partial \Gamma_{RED}}{\partial t} \qquad (7\text{-}13)$$

which is to be combined with Eq. (7-5) or (7-7), depending on reversibility.

In order to obtain a mathematical solution for the faradaic current, it is only necessary to know the flux at the electrode and to multiply it by nFA, as in Eq. (7-9). The value of the flux cannot, however, be determined without introducing additional information about the bulk of the solution and about the initial conditions. The differential equations of Fick's second law can provide the needed information:

$$\frac{\partial C_{OX}}{\partial t} = D_{OX}\left(\frac{\partial^2 C_{OX}}{\partial x^2}\right) \qquad (7\text{-}14)$$

$$\frac{\partial C_{RED}}{\partial t} = D_{RED}\left(\frac{\partial^2 C_{RED}}{\partial x^2}\right) \qquad (7\text{-}15)$$

The initial state of the system is given by:

$$\left.\begin{array}{l} C_{OX} = C_{OX}^* \\ C_{RED} = C_{RED}^* \end{array}\right\} (\text{all } x, t = 0) \qquad (7\text{-}16)$$

The assumption of semi-infinite diffusion provides two additional boundary conditions:

$$\left.\begin{array}{l} C_{OX} = C_{OX}^* \\ C_{RED} = C_{RED}^* \end{array}\right\} (x \rightarrow \infty, \text{ all } t) \qquad (7\text{-}17)$$

where the asterisks indicate bulk concentrations. "Infinite distance," for LSV, represents perhaps one millimeter.

The left sides of Eqs. (7-14) and (7-15) represent the time variation of the concentrations. Modifications must be made if homogeneous chemical reactions occur that involve either OX or RED. For example, if RED is involved in a first-order reaction that generates an electrochemically inactive product, the equations become:

$$\frac{\partial C_{OX}}{\partial t} = D_{OX}\left(\frac{\partial^2 C_{OX}}{\partial x^2}\right) \qquad (7\text{-}18)$$

$$\frac{\partial C_{RED}}{\partial t} = D_{RED}\left(\frac{\partial^2 C_{RED}}{\partial x^2}\right) - k_{hom} C_{RED} \qquad (7\text{-}19)$$

which reflect the effect of the homogeneous reaction rate constant k_{hom}. The above equations describe phenomena occurring within the diffusion layer.

The complete mathematical treatment involves two surface conditions, such as Eqs. (7-5) and (7-10), and two differential equations, such as Eqs. (7-14) and (7-15), as well as the accompanying boundary conditions. The solution is not simple, because of the exponential dependence on time[†] for both reversible and irreversible systems. The original work of Randles and Ševčík [1, 2] permits the calculation of the peak current i_{pk} by means of their well-known equation:

$$i_{pk} = 269 \, n^{3/2} \, AD^{1/2} \, v^{1/2} \, C_{OX}^*$$ (7-20)

where a 3/2-power dependence on the number of electrons is involved. For the numerical constant given, the diffusion coefficient must be in $cm^2 s^{-1}$, the area in cm^2, and the concentration in moles/liter. The equation can be used to calculate the concentrations when all other quantities are known, but analytically it is preferable to follow a calibration procedure. Good accuracy is obtained over the range 10^{-3} to 10^{-6} M [8].

The expression for the entire voltammetric curve, as opposed to the value at the peak, is more difficult to obtain, but several numerical calculations have been published, including the classical paper by Nicholson and Shain [3], from which much of the present discussion is taken.

Their expression for the reversible case consists of an explicit portion and a tabulated *current function*, χ_{rev}:

$$I = 602 \, n^{3/2} \, AD^{1/2} \, v^{1/2} \, C_{OX}^* \, \chi_{rev}$$ (7-21)

where we have replaced the symbol for the Nicholson-Shain current function $[\pi\chi(at)]$ by χ_{rev}. A shortened tabulation of values of χ_{rev} is given in Table 7-1. For more details, consult [3]. The symbol $E_{1/2}$ is here defined as $E^{\circ\prime} + (0.0592/n)$ log $(D_{red}/D_{ox})^{1/2}$, just as in polarography. The table shows the peak to occur at 28.5 mV beyond $E_{1/2}$ from which we may deduce that the anodic and cathodic peaks for the same reversible couple are separated by $(2 \times 28.5/n) = 57/n$ mV.

At the half-wave potential, the function χ_{rev} has the value 0.38, which is 85 percent of the peak value, 0.446, and thus corresponds to 85 percent of i_{pk}. This suggests a convenient way of calculating $E_{1/2}$ from an LSV curve (cf. Figure 7-2).

Note that the half-wave potential for voltammetry at a fixed-area electrode cannot be defined as the potential where the current is half its maximum value. Nevertheless, in both polarography and LSV, the half-wave potential is a fundamental point of the curve. In LSV, however, the curve is asymmetric, giving a peak, as seen in Figure 7-2.

Another useful parameter of the curve is the half-peak potential, $E_{pk/2}$. This can be measured with precision, whereas the peak potential E_{pk} itself is difficult

---

[†]Considerable mathematical simplification might be obtained by scanning the voltage logarithmically rather than linearly, using a function of the form $E = X + Y \log t$, where X and Y are constants. In this case, Eq. (7-6) would reduce to:

$$C_{OX}^s/C_{RED}^s = (\text{constant}) \cdot (1/t)$$ (7-19a)

TABLE 7-1
Values of χ_{rev} at Various Potentials
(Relative to $E_{1/2}$)

$n(E - E_{1/2})$ (mV)	χ_{rev}	$n(E - E_{1/2})$ (mV)	χ_{rev}
+120	0.01	−10	0.42
100	0.02	−20	0.44
80	0.04	−28.5[a]	0.446
60	0.08	−30	0.44
50	0.12	−40	0.44
40	0.16	−50	0.42
30	0.21	−60	0.40
20	0.27	−80	0.35
10	0.33	−100	0.31
0	0.38	−120	0.28

[a]maximum.

to measure because of the typically broad summit. E_{pk} can be calculated from $E_{pk/2}$ by the relation:

$$E_{pk} = E_{pk/2} \pm \frac{56.5}{n} \text{ mV} \qquad (7\text{-}22)$$

The negative sign applies to anodic peaks. This equation is sometimes written in the equivalent form:

$$E_{pk} = E_{pk/2} \pm 2.2 \frac{RT}{nF} \text{ mV} \qquad (7\text{-}23)$$

Consult Figure 7-2 for a graphical relation between these quantities.

In the case of a cyclic voltammogram, the peaks are separated by exactly $57/n$

TABLE 7-2
Peak Separation for a One-Electron
Process [3] (mV)

$E_f - E_{pk}$	$E_{pk(anod)} - E_{pk(cath)}$
75	61.8
100	60.5
200	58.3
300	57.8
(Limit)	57.0

mV only if the forward scan caused a sufficiently extensive change from OX to RED. This can be ensured by using a wide potential sweep, or preferably by pausing at E_f for a minute or so before starting the reverse scan. In case neither is done, the reverse peak will be slightly shifted as shown in Table 7-2. From these data, it appears than $59/n$ is probably a better estimate than the theoretical $57/n$ for reversible peak separation.

Irreversible Processes

If the electron-transfer process is slow, the differential equations contain the heterogeneous rate constant $k°$ and the transfer coefficient α. The solution of the equations involving this condition, as given by Nicholson and Shain [3], is:

$$I = 602 \, n(\alpha n_\alpha)^{1/2} \, AD^{1/2} \, v^{1/2} \, C_{OX}^* \, \chi_{irrev} \qquad (7\text{-}24)$$

where again a reduction is assumed. Note the unusual replacement of $n^{3/2}$ by $n(\alpha n_\alpha)^{1/2}$, where n_α denotes the number of electrons involved in the charge-transfer step, a rather inaccessible quantity, difficult to separate from α experimentally.

The values of χ_{irrev} are given by Nicholson and Shain [3].[†] The maximum of the curve occurs at a value $\chi_{irrev} = 0.496$. If we assume α to be 0.5, we can calculate the ratio of reversible to irreversible peaks to be:

$$\frac{i_{pk\,(rev)}}{i_{pk\,(irrev)}} = \frac{0.446}{\sqrt{0.5} \times 0.496} = 1.27 \qquad (7\text{-}25)$$

so that, everything else being constant, a reversible peak is some 30 percent higher than an irreversible one.

The irreversible peak current suffers displacement away from $E_{1/2}$ that result from two factors: small values of both $k°$ and of α. The basic formula describing this displacement is [9]:

$$E_{pk} = E^{°\prime} - \frac{RT}{\alpha n_\alpha F}\left(0.77 \ln \sqrt{Db} - \ln k° + \frac{\ln b}{2}\right) \qquad (7\text{-}26)$$

where $b = \alpha n_\alpha Fv/RT$.

By using this equation it is possible to calculate the kinetic parameters αn_α and $k°$. A careful examination of the equation indicates that the contribution of $k°$ to the peak displacement can be compensated by appropriate changes in the scan rate v. Thus, no matter how large $k°$ may be, that is, no matter how reversible the reaction may be, one can make it behave irreversibly by choosing a sufficiently fast scan rate. Conversely, any reaction, at least in principle, can be made reversible by sweeping slowly enough.

[†]They use the notation $[\pi\chi(bt)]$.

These comments underline the fact that electrochemical reversibility is a property of the experimental technique as much as of the chemical system involved.

Coupled Chemical Reactions

If either species OX or RED is involved in a side reaction, an appropriate contribution to the differential equations must be included. The results in each case contain the rate constants of the chemical reactions, and, in addition, k° and αn_α if the process is irreversible. A few of the many possible cases are the following:

$$Z \underset{}{\overset{k}{\rightleftharpoons}} OX + ne^- \longrightarrow RED \quad (CE) \tag{7-27}$$

$$OX + ne^- \longrightarrow RED \underset{}{\overset{k}{\rightleftharpoons}} Z \quad (EC) \tag{7-28}$$

$$Z_1 \underset{}{\overset{k_1}{\rightleftharpoons}} OX + ne^- \longrightarrow RED \underset{}{\overset{k_2}{\rightleftharpoons}} Z_2 \quad (ECE) \tag{7-29}$$

The symbols on the right indicate the sequence of chemical and electrochemical steps involved. Such reactions have been thoroughly investigated [3-5, 10-13], but the details are beyond the scope of our discussion.

The measurement of i_{pk} necessitates a knowledge of the baseline. The problem is especially important on the return scan. If the mechanistic requirements permit, it might be advisable to pause at the end of the forward scan until the current decays to a low value. The effect is illustrated in Figure 7-6. A less desirable alternative is to make a separate experiment recording the current as a function of time

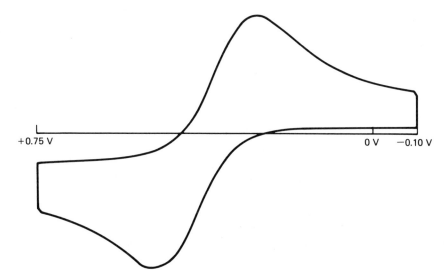

+0.75 V   0 V   −0.10 V

Figure 7-6. A cyclic voltammogram, showing the drop in current caused by a pause at −0.10 V and again at +0.75 V.

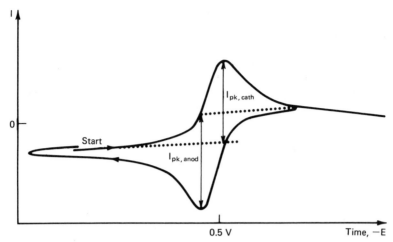

Figure 7-7.   A possible method of correction for base line. At E_f the scan stops but recording continues at constant potential.

after stopping the voltage sweep, as shown in Figure 7-7. The mirror image of this curve (dotted line) is then taken as the baseline for the reverse scan.

Staircase LSV

A variant of LSV, which is particularly appropriate for computer control, is staircase voltammetry, wherein the potential is advanced in steps of δE volts every θ seconds. This corresponds to a voltage ramp of $\delta E/\theta$, the equivalent of the quantity v. Any desired value of the ratio $\delta E/\theta$ can be implemented by various combinations of the two variables [16]. This is a complication in analytical work, but can be useful for mechanistic studies [15]. The sensitivity is good ($10^{-7} M$) due to the elimination of the charging current [16].

AC and Pulse LSV

The combination of LSV with AC or pulse signals offers the usual advantages of modulation. The AC method [17] has received attention both in the fundamental and second-harmonic modes [18]. The results for reversible reactions are somewhat surprising, since, at least for slow scans, the process turns out to be time-independent. On the other hand, irreversible systems give time-dependent results that can serve to elucidate kinetic parameters. Pulse voltammetry, especially differential [19], also gives unique curves. The method seems to promise good analytical results.

REFERENCES

1. A. Ševčík, *Collect. Czech. Chem. Commun.*, **1948**, *13*, 349.

2. J. E. B. Randles, *Trans. Faraday Soc.*, **1948**, *44*, 327.

3. R. S. Nicholson and I. Shain, *Anal. Chem.*, **1964**, *36*, 706.

4. R. S. Nicholson, *Anal. Chem.*, **1965**, *37*, 1351.

5. R. S. Nicholson, *Anal. Chem.*, **1965**, *37*, 667.

6. W. F. Gutknecht and S. P. Perone, *Anal. Chem.*, **1970**, *42*, 906.

7. P. Delahay, *J. Phys. Chem.*, **1966**, *70*, 2373.

8. J. W. Ross, R. D. DeMars and I. Shain, *Anal. Chem.*, **1956**, *28*, 1768.

9. P. Delahay, *J. Am. Chem. Soc.*, **1953**, *75*, 1190.

10. R. H. Wopschall and I. Shain, *Anal. Chem.*, **1967**, *39*, 151.

11. D. S. Polcyn and I. Shain, *Anal. Chem.*, **1966**, *38*, 370.

12. G. Ginzburg, *Anal. Chem.*, **1978**, *50*, 375.

13. R. S. Nicholson, *Anal. Chem.*, **1965**, *37*, 1406.

14. J. J. Zipper and S. P. Perone, *Anal. Chem.*, **1973**, *45*, 452.

15. M. D. Ryan, *J. Electroanal. Chem.*, **1977**, *79*, 105.

16. L. H. L. Miaw, P. A. Boudreau, M. A. Pichler and S. P. Perone, *Anal. Chem.*, **1978**, *48*, 1988.

17. W. L. Unterkoffler and I. Shain, *Anal. Chem.*, **1965**, *37*, 218.

18. A. M. Bond, R. J. O'Halloran, I. Ružić and D. E. Smith, *Anal. Chem.*, **1976**, *48*, 872.

19. K. F. Drake, R. P. Van Duyne and A. M. Bond, *J. Electroanal. Chem.*, **1978**, *89*, 231.

Chapter 8

VOLTAMMETRY:
V. FINITE DIFFUSION

The majority of electroanalytical methods utilize a relatively large volume of solution, extending to some distance from the electrode. If diffusion is the controlling transport mechanism, a concentration profile will be formed in the solution as the reaction proceeds. As seen previously, the thickness of the diffusion layer is well below 1 mm. Because of the thinness of this layer, the walls of the cell can be considered to be at infinity. This mode of transport is called *semi-infinite diffusion*. (The "semi" refers to the fact that the process extends in one direction only.)

Another type of behavior occurs when the diffusion is limited rather than infinite. In this event, the response is quite different, even though the excitation may be conventional. The most important possibilities [1] are: (1) The reagent is confined to a layer of solution limited by an inert wall or by another electrode. This can be called *thin-layer bounded diffusion*. It is implemented with so-called thin-layer cells. (2) The diffusion may be confined to a porous membrane in contact with a stirred solution. This is *thin-layer nonbounded diffusion*. Finally, (3), the reagent may be immobilized by adsorption or binding at the electrode surface, in which case only very slight diffusion (or none at all) is needed for the reaction to proceed. This can be called an *immobilized reagent*.

THIN-LAYER CELLS

Although electrolysis with microliter volumes had been attempted as early as 1955 [2], only recently [3, 4] has a thorough investigation been made of electrolysis under conditions where the thickness of the cell is smaller than the diffusion layer. The exact shape of the concentration profile varies with the type of potential-time curve used. In order to compare the infinite and bounded diffusion cases, let us first consider the application of a large voltage step to a cell with semi-infinite diffusion. This is a Cottrell-type procedure, and it can be shown that the concen-

130

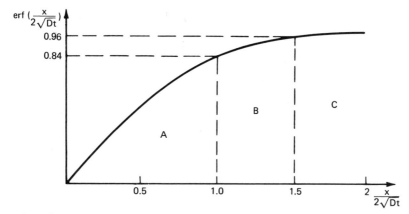

Figure 8-1. Graph of the error function. The argument of "erf" represents the distance into the solution, in units of $2\sqrt{Dt}$ cm.

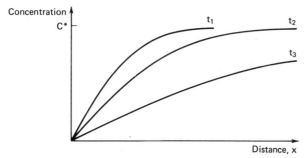

Figure 8-2. An alternative plot of Eq. (8-1) as concentration profiles at various times. Curves 1, 2, and 3 correspond to increasing time intervals. The initial slopes of the curves are given by $(dC/dx)^s$, and, if plotted against t, produce the Cottrell relation.

tration profile follows a functional dependence of the type:

$$C = C^* \text{ erf } [x/(2\sqrt{Dt})] \tag{8-1}$$

where "erf" is a tabulated function, the *error function*.†

The concentration depends on both the distance x from the electrode and the time t. A graph of the spatial dependence at a specified time, is shown in Figure 8-1. As the time increases, the concentration profile becomes shallower, as seen in Figure 8-2. In Figure 8-1, three regions can be distinguished. In region A the profile is nearly linear; if the time and thickness of the cell are such that the function falls within this region, we speak of a *thin-layer cell*. In this case the diffusion transport involves the entire volume of the cell, and the mathematical equations describing the relation between current and voltage are relatively simple.

†It would be a mistake to believe that the error function, in this context, has any relation to errors.

Region A corresponds to arguments of the error function less than unity, or:

$$x_d < 2\sqrt{Dt} \qquad (8\text{-}2)$$

Region B corresponds to a more complicated situation and is of little analytical interest. Region C, corresponds to arguments larger than about 2; this can be described by the condition:

$$x_d > 4\sqrt{Dt} \qquad (8\text{-}3)$$

Beyond this point, the solution is practically untouched by diffusion, and the behavior can be approximated by semi-infinite treatment. The discussion above is only qualitative in nature, since the presence of the outer boundary influences the concentration profile, due to the zero flux of matter at the outer wall.

Equation (8-1) depends on time, so that for a long enough period, any cell will behave like a thin-layer cell, and correspondingly, for short enough times, any cell will obey semi-infinite mathematics. As a rule, the electrodes must be separated by less than 0.1 mm for thin-layer operation to prevail.

There are many ways of implementing such thin electrolyte layers, a few of which are shown in Figure 8-3. In cases a and b, the current must travel the length of a slice of solution, and resistance effects may become important. In c, this is eliminated by the use of a porous material in contact with the electrolyte. This latter case is more properly discussed in the next section, as a membrane electrode.

The fundamental characteristic of thin-layer operation is that the whole cell is essentially in equilibrium with the electrode. In other words, the quantities C^* and C^s are practically equal at all times. For this to be true, the cell must be operated in region A of Figure 8-1. For usual cells, this requires experimental durations longer than about 1 s. Consequently, any potentiostatic operation, such as LSV, can be utilized if the scan is sufficiently slow.

It follows that the concentration ratio in the entire cell can be approximated by:

$$\frac{C_{OX}}{C_{RED}} = \exp\left\{\frac{nF}{RT}(E - E^{\circ\prime})\right\} = \Theta \qquad (8\text{-}4)$$

which is a form of the Nernst equation. In this closed system, unlike LSV, $C_{ox} + C_{red} = C_{total}$, and thus:

$$C_{OX} = C_{total}\left(1 - \frac{1}{1 + \Theta}\right) \qquad (8\text{-}5)$$

as the reader can easily ascertain. It is now possible to calculate the current as a function of time, since it is related to the chemical change by the factor nF. The total amount of the substance is Ax_dC, where A is the area and x_d the thickness. This corresponds to $Q = nFAx_dC$ coulombs, and the current is given by the

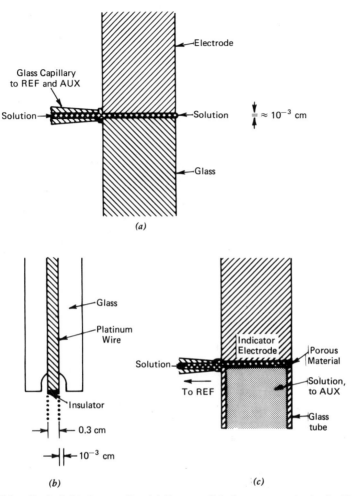

Figure 8-3. Typical thin-layer cells. (a) Two parallel plates, one or both of which are electrodes; the plates can be positioned by micrometer mechanism. (b) Cylindrical cavity probe immersed in the solution. (c) Absorbent cell.

derivative:

$$\frac{dQ}{dt} = I = nFAx_d \cdot \frac{dC_{OX}}{dt} \tag{8-6}$$

The derivative can be obtained from Eqs. (8-4) and (8-5), to give:

$$I = \frac{n^2 F^2 Ax_d vC_{total}}{RT} \cdot \frac{\Theta}{(1 + \Theta)^2} \tag{8-7}$$

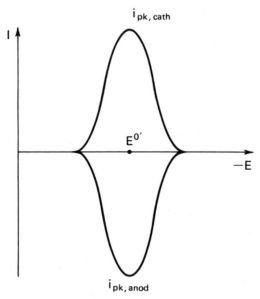

Figure 8-4.   A thin-layer cyclic voltammogram.

where v is the scan rate, dE/dt. This is a reasonably simple equation as compared with that of LSV. It corresponds to a peaked curve, with the anodic and cathodic maxima at the same potential, $E^{\circ\prime}$ (Figure 8-4). The current at the peak, where $E = E^{\circ\prime}$ and $\Theta = 1$, is given by:

$$i_{pk} = \frac{n^2 F^2 A x_d v C_{total}}{4RT} \qquad (8\text{-}8)$$

This behavior is different from conventional LSV, where there is a voltage shift between the anodic and cathodic peaks. The appearance of a shift in thin-layer operation would indicate that not all the solution is in equilibrium with the electrode. This can occur if the sweep rate is too fast (greater than about 10 mV/s), or in case of sluggish kinetics. This permits the use of thin-layer cells for kinetic measurements [3]. For example, the transfer coefficient can easily be obtained from the peak heights:

$$\alpha = \frac{i_{pk(cath)}}{i_{pk(cath)} - i_{pk(anod)}} \qquad (8\text{-}9)$$

Correspondingly simple relationships exist for the heterogeneous rate constant, which causes peak separation of about 120 mV per order of magnitude of k°.

Galvanostatic operation of thin-layer cells is also of interest. If a constant current is applied, a given species will be electrolyzed [6] in the *transition time*:

$$\tau = \frac{nF}{I} \cdot A x_d C \qquad (8\text{-}10)$$

where the product Ax_dC gives the number of moles present. In practice, the time τ is somewhat shorter than given by this equation, since as the electrolysis proceeds, at some point there will be an insufficient amount of active substance to support the current, and the electrolysis of some other species will take place. The effect is more pronounced at higher currents and with thicker cells.

Coulometry, while particularly easy to apply in thin-layer cells, exhibits a sensitivity of only about 10^{-5} M. Note, however, that the sensitivity in total number of moles is outstanding, due to the small volume involved (typically 10 μL or less), which means that only a fraction of a nanomole is reacted.

Even more advantageous is the use of pulse polarography [8], where the detection limit is about 10^{-7} M, using the same volume, corresponding to only a picomole of reagent.

Many electrochemical methods involving semi-infinite diffusion can be adapted to thin-layer operation, often resulting in simpler mathematical modeling. The exhaustive nature of the process and the small volume of solution cause the ratio of active species to adsorbed impurity to be advantageous. This permits the study of monolayers [10] and adsorption effects [9] with much better experimental control than available with conventional approaches.

ELECTROCHEMISTRY WITH IMMOBILIZED REAGENTS

The next logical step beyond thin-layer cells involves actually binding the active species to the surface of the electrode, which is tantamount to using a very thin cell. The process of immobilization can be affected by adsorption, by chemisorption [9], or by coating the electrode with a substance chemically bound to it [11]. The amount of substance available is given by the surface excess, Γ, previously defined. Normally the electrode is loaded with reagent in one solution and transferred to another for use. Under these conditions the surface excess represents the totality of substance involved in the redox process, which for an area A becomes the product $A\Gamma$. The concept of concentration is no longer valid in the conventional sense, and one must substitute the activity associated with a given surface excess:

$$\alpha_{OX} = \gamma_{OX}\Gamma_{OX} \qquad (8\text{-}11)$$

and

$$\alpha_{RED} = \gamma_{RED}\Gamma_{RED} \qquad (8\text{-}12)$$

where the quantities γ_{ox} and γ_{red} are proportionality constants with the units of cm^{-1}.

The behavior of such immobilized substances depends on the status of both OX and RED. If only one of them is strongly held, the system follows a mixture of bounded and unbounded diffusion kinetics; if, however, both species are immobilized, the behavior is quite similar to that of a thin-layer cell. For LSV [9] the current is given in this latter case by:

$$I = 4_{i\mathrm{pk}} \cdot \frac{\Theta}{(1 + \Theta)^2} \tag{8-13}$$

This should be compared with Eqs. (8-7) and (8-8).

The constant I_{pk} represents the equal heights of the anodic and cathodic peaks, which occur at the same potential for reversible reactions. The behavior of immobilized reagents has been amply documented both for adsorbed species and for chemically bound materials, the so-called *chemically modified electrodes* [11]. Shifts in the potential of the peaks occur if $\gamma_{\mathrm{ox}} \neq \gamma_{\mathrm{red}}$, and also if the electron-transfer reaction is slow, as seen in thin-layer cells.

A somewhat different dimension is provided by the use of surface modifiers to alter the catalytic properties of an electrode surface. Thus if a reversibly reducible species is chemically bound in the form of a polymer layer [12], the electron transfer can occur in two steps:

$$(\mathrm{POL})_{\mathrm{OX}} + e^- \longrightarrow (\mathrm{POL})_{\mathrm{RED}} \tag{8-14}$$

$$(\mathrm{POL})_{\mathrm{RED}} + \mathrm{OX} \longrightarrow (\mathrm{POL})_{\mathrm{OX}} + \mathrm{RED} \tag{8-15}$$

so that the net reaction, $\mathrm{OX} + e^- \longrightarrow \mathrm{RED}$, is controlled by the rate of reaction (8-14). The entity POL could be, for example, poly-(*p*-nitrostyrene) or polyvinyl-ferrocene. The choices of active species and the modes of anchoring to the substrate are vast. Consequently one can foresee a great future in such electrodes, both in analysis and in preparative applications.

MEMBRANE ELECTRODES

We turn now to the non-bounded case of diffusion control. This occurs most commonly when the electrode is in contact with a membrane in which the diffusion is slower than in the unrestricted solution [13]. In this case, the concentration profile will be similar to that indicated in Figure 8-5, established after a short initial transient.

The membrane can be considered as a porous material filled with the solution (in other words, there is only mechanical hindrance). The diffusion coefficient can be assumed to be basically unchanged by the presence of the membrane, and only modified by the fact that transport cannot occur in a straight path:

$$D_{\mathrm{effective}} = D/q \tag{8-16}$$

where q is a dimensionless number, the *tortuosity factor* [14]. In addition, the electrode area must be corrected by a factor ϵ, the porosity of the membrane, defined as the ratio of the area open to transport to the total area.

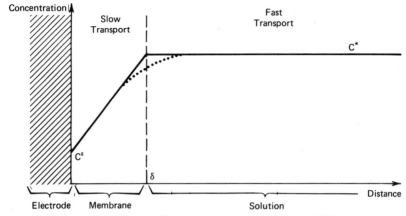

Figure 8-5. The concentration profile in a membrane electrode. Full line: simple model; dotted line: actual profile.

The response of a membrane electrode of thickness δ, changes in several stages following the application of a potential step. Initially, it is time-dependent, and semi-infinite diffusion prevails. After a period of about

$$t_1 = 0.2 \frac{\delta^2}{D/q} \tag{8-17}$$

the process converts to a finite diffusion transient [13]. At still later times, this gives place to steady-state diffusion, which occurs at about

$$t_2 = 2 \frac{\delta^2}{D/q} \tag{8-18}$$

It follows that one can operate a membrane electrode under semi-infinite or under thin-layer nonbounded diffusion depending on the time scale and the membrane thickness.

For long experiments, a steady-state current is attained, given by:

$$I = nF \frac{A \epsilon}{\delta} \frac{D}{q} C^* \tag{8-19}$$

This relation can serve to measure the concentration C^*, and it can be employed as a convenient approach to the determination of the diffusion coefficient by comparison with a standard.

Membrane electrodes tend to be slow in their response. Thus for $\delta = 0.01$ cm, $D = 10^{-5}$ cm^2 s^{-1} and $q = 10$, a steady state is attained only after about 200 seconds. The presence of the membrane has a favorable isolating effect in solutions with solid suspensions, such as many biological fluids, where membrane electrodes

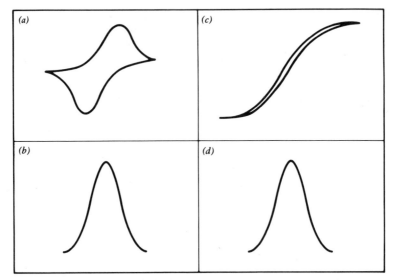

Figure 8-6.   Response of an ultramicro electrode to linear sweep voltammetry (*c*) and to DPP (*d*), compared with a conventional electrode in the same solution (*a* and *b*).

have found some use [15].   An example is *Clark oxygen sensor*, which employs a gas-permeable membrane.   The electrode is highly selective, responding only to oxygen.

ULTRAMICROELECTRODES

Another special case of diffusion occurs when the area of the electrode is very small ($<10^{-4}$ cm^2). Taking into account that the diffusion layer reaches this dimension after less than a millisecond, it can be seen that, as a result, the transport of active species is relatively constant. A linear sweep voltammogram generates a wave rather than a peak, as seen in Figure 8-6 [16, 17].

REFERENCES

1.   H. E. Keller and W. H. Reinmuth, *Anal. Chem.*, **1972**, *44*, 434.

2.   R. Schreiber and W. D. Cooke, *Anal. Chem.*, **1955**, *27*, 1475.

3.   A. T. Hubbard and F. C. Anson, in "Electroanalytical Chemistry: a Series of Advances" (A. J. Bard, ed.), Vol. 4, p. 129, Dekker, New York, **1970**.

4.   A. T. Hubbard, *CRC Crit. Rev. Anal. Chem.*, **1973**, *3*, 201.

5.   G. M. Tom and A. T. Hubbard, *Anal. Chem.*, **1971**, *43*, 671.

6.   C. R. Christensen and F. C. Anson, *Anal. Chem.*, **1963**, *35*, 205.

7.   D. M. Oglesby, L. B. Anderson, B. McDuffie, and C. N. Reilley, *Anal. Chem.*, **1965**, *37*, 1317.

8. T. P. DeAngelis, R. E. Bond, E. E. Brooks, and W. R. Heineman, *Anal. Chem.*, 1977, *49*, 1792.

9. R. F. Lane and A. T. Hubbard, *J. Phys. Chem.*, 1973, *77*, 1401.

10. E. Schmidt and H. R. Gygax, *J. Electroanal. Chem.*, 1966, *12*, 300.

11. R. W. Murray, *Acc. Chem. Res.*, 1980, *13*, 135.

12. C. P. Andrieux and J. M. Saveant, *J. Electroanal. Chem.*, 1980, *111*, 377.

13. R. C. Bowers and A. M. Wilson, *J. Am. Chem. Soc.*, 1958, *80*, 2968.

14. J. H. Clausen, G. B. Moss, and J. Jordan, *Anal. Chem.*, 1966, *38*, 1398.

15. L. C. Clark, Jr., R. Wolf, D. Granger, and Z. Taylor, *J. Appl. Physiol.*, 1953, *6*, 189.

16. M. A. Dayton, J. C. Brown, K. J. Stutts, and R. M. Wightman, *Anal. Chem.*, 1980, *52*, 946.

17. J.-L. Ponchon, R. Cespuglio, F. Gonon, M. Jouvet, and J.-F. Pujol, *Anal. Chem.*, 1979, *51*, 1483.

Chapter 9

CONTROLLED-CURRENT METHODS

In the last several chapters we have considered electrochemical methods that are based on the application of a programmed potential to the cell with concomitant observation of the resulting current. Now we turn to the reverse procedure, wherein a current conforming to a prescribed model is forced to flow through the cell, and the voltage appearing at the working electrode is monitored.

We will be concerned with current-step procedures, in which a current is abruptly turned on, and the voltage response recorded. In one of these, *chronopotentiometry*, the current is held constant or programmed to follow a selected function of time, and the changing potential is observed.

In another method, called *coulostatic analysis*, the time-integral of the current, rather than the current itself is controlled. Again, the potential is recorded against time.

CHRONOPOTENTIOMETRY

The Basic Experiment

Conventionally, the chronopotentiometric experiment is carried out with a constant current of several milliamperes. A three-electrode cell is connected to a galvanostat and provided with a stationary working electrode, an isolated platinum counter electrode, and a reference element in a Luggin capillary, as shown in Figure 9-1.

Let us assume that the cell contains a small concentration of reducible species, say 10^{-4} M Pb^{++}, in a supporting electrolyte. Prior to time zero (the start of the electrolysis), the mercury electrode senses the concentration of Pb^{++} ion, but this cannot establish a valid potential in the absence of reduced lead.† As soon as the

---

†This potential is actually determined by some other process, such as $Pb^{++} + 2Hg \longrightarrow Hg_2^{++} + Pb$.

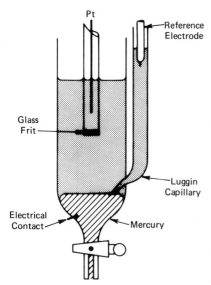

Pt

Reference
Electrode

Glass
Frit

Luggin
Capillary

Electrical
Contact

Mercury

Figure 9-1. A typical cell for chronopotentiometry. The Luggin capillary, with its orifice close to the mercury surface, minimizes the effect of resistive voltage drop within the solution.

current starts to flow, however, lead atoms begin to accumulate in the mercury, establishing the ratio of Pb^{++} to Pb in mercury. Its value, initially very large, decreases with time, so that the potential shifts towards less positive values. The relative change is rapid during the first few seconds, then becomes more gradual. Throughout the entire process, the faradaic current must be supported by the diffusion of the Pb^{++} ions from the bulk of the solution to the electrode surface. The ratio eventually becomes much less than unity, and the potential becomes more and more negative, until some other component of the solution is reduced and contributes to the support of the current. These relations are illustrated in Figure 9-2.

Repetition of this basic experiment with increasing concentrations of Pb^{++} ion will give similar curves, but with longer time intervals (*transition times*) between the initial and final sharply rising portions of the curve. Measurements show the transition time, τ, to be proportional to the *square* of the concentration, and we can write:

$$C_{OX}^* = k\sqrt{\tau} \qquad (9\text{-}1)$$

where k is a proportionality constant.

In general, chronopotentiograms are less well-defined than that shown in Figure 9-2, principally because of double-layer effects. The rising portions of the curve are often far from vertical and not parallel to each other, as seen in Figure 9-3. This necessitates an empirical approach to the measurement of transition times from the graph. Several geometrical techniques by which to do this have been proposed. The best of them [1], illustrated in the figure, is to draw tangents to the three segments of the curve and to consider τ to be the horizontal distance between the two intersections.

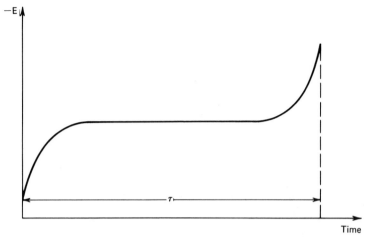

Figure 9-2.   An idealized chronopotentiogram, without changing current. The transition time, denoted by τ, is proportional to the square of the concentration.

Similar experiments with two reducible species present result in curves with two plateaus, and hence two transition times (Figure 9-4). These are not directly additive, however. If both reducible species are cations, such as Pb^{++} and Cd^{++}, of equal charge and equal concentrations, the transition times will, nevertheless, be different; in fact, under these conditions, $\tau_{Cd} = 3\tau_{Pb}$.

Theoretical Considerations

As seen previously, the Cottrell equation gives the current when the potential is held constant in a diffusion-controlled situation. Now we require a comparable

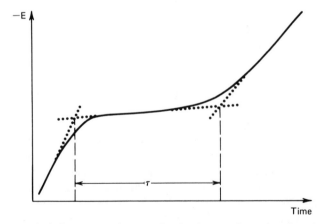

Figure 9-3.   A typical chronopotentiogram, showing how to determine the value of the transition time.

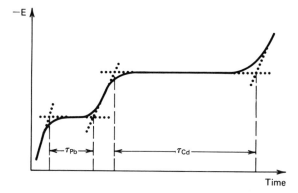

Figure 9-4. A chronopotentiogram for a mixture of equal concentrations of lead and cadmium.

relation for the potential at constant current, and this can be derived in the same general way. We will assume that both OX and RED are soluble, and that initially only OX is present. (The treatment given for a reductive process is equally applicable to an oxidation with reversal of appropriate signs.)

The procedure is to solve the Fick equation for the surface concentration, $C_{OX}^s(t)$, to derive a similar expression for $C_{RED}^s(t)$, and to substitute them into the Nernst equation. The required relations are:

$$C_{OX}^s(t) = C_{OX}^* - \frac{2I}{\pi^{1/2} nFAD_{OX}^{1/2}} \sqrt{t} = C_{OX}^* - W\sqrt{t} \qquad (9\text{-}2)$$

and

$$C_{RED}^s(t) = \frac{2I}{\pi^{1/2} nFAD_{RED}^{1/2}} \sqrt{t} = W\sqrt{t} \qquad (9\text{-}3)$$

where $W = 2I/(\pi^{1/2} nFAD^{1/2})$, provided that $D_{OX} = D_{RED}$. Substituting these into the Nernst equation gives:

$$E = E^\circ + \frac{RT}{nF} \ln \left(\frac{D_{RED}}{D_{OX}}\right)^{1/2} + \frac{RT}{nF} \ln \left(\frac{C_{OX} - W\sqrt{t}}{W\sqrt{t}}\right) \qquad (9\text{-}4)$$

It will be noted that the first two terms of Eq. (9-4) fit exactly the definition of the polarographic half-wave potential, but we will use the symbol $E_{\tau/4}$ to stress its different experimental significance (of which more later):

$$E = E_{\tau/4} + \frac{RT}{nF} \ln \frac{C_{OX}^* - W\sqrt{t}}{W\sqrt{t}} \qquad (9\text{-}5)$$

The end of the transition time corresponds to the point where the log term becomes infinite,[†] namely where $C_{OX}^* = W\sqrt{t}$. Hence, we can write τ in place of t:

$$C_{OX}^* = W\sqrt{\tau} \tag{9-6}$$

and introduce this into Eq. (9-5) to give:

$$E = E_{\tau/4} + \frac{RT}{nF} \ln \frac{\tau^{1/2} - t^{1/2}}{t^{1/2}} \tag{9-7}$$

This relation is called the *Karaoglanoff equation*,[‡] and corresponds to the curve in Figure 9-2.

The logarithmic term in Eq. (9-8) disappears when the argument is unity, which occurs if

$$t = \tau/4 \tag{9-8}$$

Therefore, when the time is equal to one-fourth of the transition time, $E = E_{1/2} = E_{\tau/4}$.

An expression relating the transition time to the concentration of OX can be obtained by combining Eqs. (9-4) and (9-7):

$$\tau^{1/2} = \frac{\pi^{1/2} nFAD_{OX}^{1/2}}{2I} \cdot C_{OX}^* \tag{9-9}$$

This is the *Sand equation*.

It can be seen from this relation that the quantity $I\tau^{1/2}$ is proportional to concentration but independent of the current. Hence, a plot of this quantity against the applied current, I (Figure 9-5), consists of a horizontal straight line for any concentration (assuming that the electrode area is unchanging). This provides a convenient criterion by which to establish whether or not diffusion control is actually operating.

When two electroactive substances are present, the curve contains two transition periods. In common with polarography, as the second species begins to be reduced, the first continues to react. In the present case, however, the two transition times

---

[†] Actually, of course, the potential does not increase without limit, but only until it becomes negative enough to reduce some other component of the solution; thus the *chemistry* of the system does not tend to drive the potential to infinity, such a tendency is purely mathematical.
[‡] Note the similarity of this expression to the Heyrovský–Ilkovič equation of polarography:

$$E = E_{1/2} + \frac{RT}{nF} \ln \left(\frac{I_d - I}{I} \right) \tag{9-7a}$$

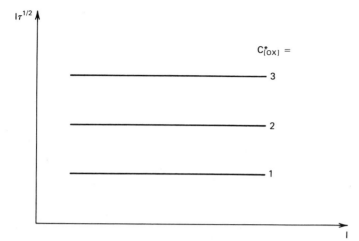

$C^*_{[OX]} =$

Figure 9-5. Illustrating the independence of the $I\tau^{1/2}$ product with respect to the current for a diffusion-limited system.

affect each other in a nonlinear manner. The governing relation becomes [2, 3] :

$$(\tau_1 + \tau_2)^{1/2} - \tau_1^{1/2} = \frac{\pi^{1/2} n_2 F A D_{OX(2)}^{1/2}}{2I} \cdot C^*_{OX(2)} \qquad (9\text{-}10)$$

where the subscripts refer to the first and second species. According to Eq. (9-9), we can write $\tau_1^{1/2} = kC_1^*$, and substitute it into Eq. (9-10), and assuming that the respective values of n and D are equal, we can readily show that $\tau_2 = 3\tau_1$, in accordance with Figure 9-4.

Chronopotentiometry, in the basic form outlined above, is useful only for relatively concentrated solutions, greater than about 10^{-3} molar, hence is seldom used for analytical purposes. The fundamental weakness lies in the inherent difficulty of distinguishing between the component of the current responsible for faradaic effects and that utilized in charging the double-layer capacitance.

The rate of double-layer charging is a function of the derivative dE/dt, and since the potential changes rapidly both at the start and at the finish of the transition period, it is in those regions that charging must be dealt with.

The method of measuring τ shown in Figure 9-3 helps to diminish the effects of charging, but is limited by the accuracy with which one can draw the requisite tangents. It is possible to generate the derivative curve, obtaining two maxima, and to take the distance between them as the transition time, but this shows little improvement.

Peters and Burden [4] have shown that the minimum between these two derivative maxima occurs at a time corresponding to $\frac{4}{9}$ of τ (Figure 9-6), and that the value of the transition time can be determined with greater precision by measuring the value of the function dE/dt at the minimum point. The governing relation

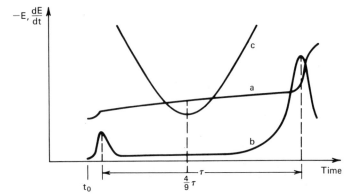

Figure 9-6.   (*a*) A chronopotentiogram for the reduction of Tl(I) in 0.1 *M* KNO$_3$ at a mercury pool cathode (*I* = 0.05 mA).   (*b*) The first derivative of curve (*a*).   (*c*) An enlarged portion of curve (*b*) around the minimum.   Vertical axis: (*a*) –*E* (volts), (*b* and *c*) *dE/dt* (volts/second). (*Analytical Chemistry [4]*, *redrawn.*)

is:

$$\tau = -\frac{27}{8} \frac{RT}{nF} \frac{1}{(dE/dt)_{\min}} \qquad (9\text{-}11)$$

This method gives results that are nearly independent of double-layer effects, since the only measurement required is taken at the point where the charging current is a minimum.

Instrumentation

The basic circuit for chronopotentiometry consists of a three-electrode cell connected to a galvanostat, with provision for recording the potential developed between the working and reference electrodes. This corresponds to the upper portion of Figure 9-7. The galvanostat is driven by a constant-current source.

Since the charging current is proportional to *dE/dt*, it is possible to correct for it electronically. Sturrock and co-workers [5] have done this by feeding a current proportional to the derivative back into the galvanostat amplifier (the lower loop of Figure 9-7) in such a way as to add to the constant current. The feedback current is adjusted to provide the needs of double-layer charging, leaving the current from the source to be used solely for faradaic purposes.†

AC Chronopotentiometry

Just as in polarography, it is possible to make use of modulation to improve the S/N ratio in chronopotentiometry. AC modulation has been reported by Bansal

†Note the formal resemblance between this charging current compensation method and the comparable technique in DC polarography.

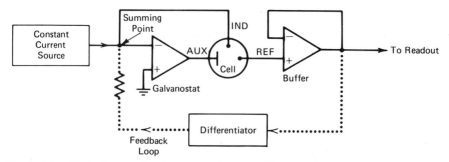

Figure 9-7. Block diagram of a chronopotentiometer. The dotted segment indicates a method of correction for charging current, whereby a signal related to the time-derivative of the potential is fed back to the summing point. This adds just enough to the constant current to compensate for the nonfaradaic component.

and Plambeck [6, 7], who applied about 3 μA at 100 Hz modulation to the DC current. The AC component of the potential at the working electrode was measured with the aid of a phase-sensitive detector (a lock-in amplifier). They worked out the theoretical equations for the time-dependence of the electrode potential and showed their applicability to reversible, quasi-reversible and irreversible processes. This method has aroused little analytical interest.

Chronopotentiometry with Increasing Current

As we have seen, the double-layer charging current is largest at the beginning and at the end of each transition time. Hence, any method that will decrease its influence at these points will improve the precision of the measurement. One way in which this can be accomplished is to use a current that varies with time.

A number of current programs have been investigated, principally those that start from near zero and increase to a relatively large value. While this diminishes the charging error at the end of the transition, when the current is large, the error at the start must be treated separately.

If no precautions are taken, the potential at the beginning of the scan is likely to assume a highly positive value (cf. Figure 9-2), which increases the initial charging transient. One way to prevent this positive excursion is to employ temporarily a potentiostat circuit at the start of an experiment, and then switch to the desired current [1]. Chow and Ewing [8] have utilized instead a constant current for a short time at the start of the electrolysis; transfer from the constant current to increasing current was made to occur at the point where the faradaic effect would be equal either way, following which, the current obeyed the relation:

$$I = bt^{1/2} \tag{9-12}$$

where b is a constant. This square-root relation brings the added advantage that transition times for multiple reductions become directly additive, as will be shown in the following equations.

Programmed current alters the functional relation of the transition time as related to bulk concentration. If the faradaic current is forced to follow the equation:

$$I = kt^q \tag{9-13}$$

where k is a constant, then it can be shown [8] that τ obeys the relation:

$$\tau^{(q + 1/2)} = \frac{\pi^{1/2} nFAD_{OX}^{1/2}}{2I} \cdot C_{OX}^* \tag{9-14}$$

As would be expected, if $q = 0$, Eq. (9-14) reverts to the equivalent form of the Sand equation [Eq. (9-9)]. On the other hand, if $q = \frac{1}{2}$, τ becomes *directly* proportional to C_{OX}^*, and furthermore, transition times for successive reducible species become additive. Thus there is substantial advantage in programming the current to vary as the square-root of time. Figure 9-8 shows representative chronopotentiograms obtained by this procedure [8]. The sensitivity of the method permits measurements down to about 10^{-6} M.

Other powers of time can be useful. For instance, $q = \frac{3}{2}$ permits increasing the current at the same rate as the area of a DME, so that a conventional chronopotentiogram results.

Murray [9] also considered the application of an exponential current:

$$I = \alpha \exp(\beta t) \tag{9-15}$$

This gave fairly good preliminary results but apparently has not been followed up.

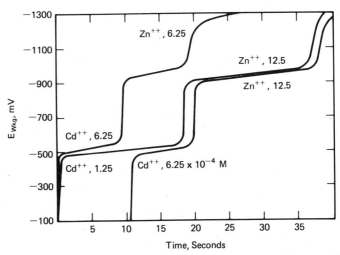

Figure 9-8. Square-root-of-time current chronopotentiograms of Cd^{++} and Zn^{++} in 0.5 M KNO_3. Switching time = 2s. (*Analytical Chemistry [9]*.)

Chronopotentiometry has not so far been applied very widely in practical analysis, though it is a useful tool in kinetic and mechanistic studies. It may well become more important in analytical applications if the double-layer effects can further be minimized by improved instrumentation.

COULOSTATIC ANALYSIS

This is an unusual technique, in that neither the potential nor the current is controlled. Rather, a particular *quantity* of electric charge is utilized. The method was first described simultaneously by Reinmuth and by Delahay [10–12].

The basic experiment can be described as follows: The cell consists of a small mercury pool, an inert counter-electrode, and a reference. A capacitor C (Figure 9-9) is first charged to potential E_{init} with the cell disconnected. The switch is then thrown to the right so that the capacitance of C is suddenly placed in parallel with the capacitance C_{dl} of the double layer at the working electrode. The charge $Q = CE_{init}$ is thus divided between C and C_{dl} at the voltage $E_{max} = Q/(C + C_{dl})$. As evident from the equivalent circuit of Figure 9-9b, the combined capacitance is

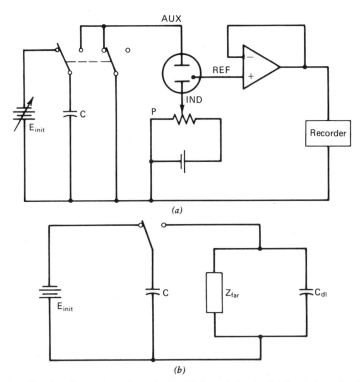

Figure 9-9. Circuitry for coulostatic analysis. (*a*) Actual circuit. (*b*) Equivalent circuit, showing the faradaic impedance and double-layer capacitance as circuit components. The solution resistance is assumed negligible.

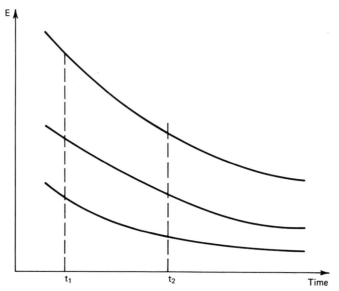

Figure 9-10.   Decay curves obtained in coulostatic analysis. The potential difference between t_2 and t_1 relates to the concentration of the reducible species.

now paralleled by the faradaic impedance and proceeds to discharge through it with a time constant that is a function of the concentration of any reducible species present. The recorder will then display a curve like those in Figure 9-10. Measurement of the change in potential over a specified interval, $(t_2 - t_1)$ in the figure, gives a convenient measure of concentration [13].

Astruc et al. [14-16] have automated the method, developing what they consider to be a new analytical technique to be designated as *discharge polarography*. Van Leeuwen [17] has reviewed critically the basic method and a number of variants. He finds its usefulness greatest in the investigations of electrode kinetics, rather than in analysis *per se*.

REFERENCES

1.   P. Bos and E. Van Dalen, *J. Electroanal. Chem.*, 1973, *45*, 165.

2.   T. Berzins and P. Delahay, *J. Am. Chem. Soc.*, 1953, *75*, 4205.

3.   P. Delahay, "New Instrumental Methods in Electrochemistry," Wiley-Interscience, N.Y., 1954, p. 191.

4.   D. G. Peters and S. L. Burden, *Anal. Chem.*, 1966, *38*, 530.

5.   P. E. Sturrock, I. L. Hughey, B. Vandreuil, G. E. O'Brien, and R. H. Gibson, *J. Electrochem. Soc.*, 1975, *122*, 1195.

6.   N. P. Bansal and J. A. Plambeck, *J. Electroanal. Chem.*, 1977, *78*, 205.

7.   N. P. Bansal and J. A. Plambeck, *J. Electroanal. Chem.*, 1977, *81*, 239.

8.   L. H. Chow and G. W. Ewing, *Anal. Chem.*, 1979, *51*, 322.

9. R. W. Murray, *Anal. Chem.*, **1963**, *35*, 1784.

10. W. H. Reinmuth, Anal. Chem., **1962**, *34*, 1159.

11. P. Delahay, *Anal. Chem.*, **1962**, *34*, 1161.

12. P. Delahay and W. H. Reinmuth, *Anal. Chem.*, **1962**, *34*, 1344.

13. P. Delahay and Y. Ide, *Anal. Chem.*, **1963**, *35*, 1119.

14. M. Astruc, J. Bonastre, and R. Royer, *J. Electroanal. Chem.*, **1972**, *34*, 211.

15. M. Astruc and J. Bonastre, *J. Electroanal. Chem.*, **1972**, *40*, 311.

16. J. Bonastre, M. Astruc, and J. L. Bentata, *Chim. Anal. (France)*, **1968**, *50*, 113.

17. H. P. Van Leeuwen, *Electrochim. Acta*, **1978**, *23*, 207.

Chapter 10

METHODS WITH CONVECTION: I. ELECTRODEPOSITION AND COULOMETRY

The dynamic methods considered in previous chapters have relied upon diffusion as their principal mode of mass transport. We will now take up a number of methods in which convection plays an essential role. One result of convective mass transport, such as by stirring, is that currents may flow that are orders of magnitude larger than those encountered in diffusion-controlled methods. This means that IR voltage drops may be much more significant, and also that Joule heating may play a part. It must be realized, however, that diffusion continues to be present; the effect of convection is to bring the electroactive species quickly to the point very close to the electrode where diffusion must finally take over.

ELECTRODEPOSITION

This is the oldest of the electroanalytical methods, as Wolcott Gibbs as early as 1864 published procedures for the determination of copper and nickel by this technique.

The Basic Experiment

Let us see how we might use electrodeposition in the analysis of a sample of a bronze that contains copper and tin with a lesser amount of lead. We will assume that the sample has already been brought into solution in a tartrate buffer at pH about 5.

Instrumentally, there are two modern options: potentiostatic control and galvanostatic control. Of these, the first is by far the more useful. (Current control is used extensively in coulometric titrimetry, treated later in this chapter, but does not in itself provide a mechanism for separation of metals from a mixed solution.)

For the present experiment, we select a three-electrode cell equipped with two large platinum electrodes and a conventional SCE reference. The cell is connected to a potentiostat so that the cathode can be maintained at any desired potential.

The first step of the analysis is to deposit copper, the most electropositive of the three metals, at -0.35 V. This step is continued until the current drops to a small fraction of its initial value. The cathode is weighed and returned to the solution for deposition of lead at -0.60 V, after which it is reweighed. This is followed by acidification to destroy the tin-tartrate complex, permitting the deposition of tin at -0.65 V.

This example shows the importance of careful control of the potential. If the cathode potential had been set initially at -0.60 V, then both Cu and Pb would have been deposited together as an alloy. If the initial potential had been -2 V, the deposit would have contained all three metals.

Unfortunately, the standard potentials, E°, as found in tables, do not provide enough information to apply directly in deciding what voltage to use to effect a particular separation. It is essential to take into account the effects of complexing agents and of pH, and this can be done by using formal potentials, $E^{\circ\prime}$. This quantity was defined in Chapter 3 and tabulated in Table 3-3 and Figure 3-4. Values of $E^{\circ\prime}$ for conditions appropriate to electrodeposition are usually unavailable, but as a guide, $E_{1/2}$ may serve the purpose.

Theoretical Considerations

Let us now examine the mathematical basis for the electrolytic separation of metallic elements. Consider a solution of a reducible species, OX, in the presence of a supporting electrolyte. The potential of the cathode is set sufficiently negative that the surface concentration is zero. The resulting current is limited by the rate at which ions of OX are transported to the surface, so we can write:

$$I(t) = -nF\frac{dM}{dt} \tag{10-1}$$

where dM represents the number of moles delivered to the electrode in time dt. Due to the stirring, the concentration is uniform throughout the solution, and is affected in its totality by the redox process at the electrode. Noting that

$$C_{OX}^* = M/V \tag{10-2}$$

(V being the volume containing M moles), we can rewrite Eq. (10-1) as:

$$I(t) = -nFV\frac{dC_{OX}^*}{dt} \tag{10-3}$$

Observe that C^* is now a variable, in contradistinction to its use in previous chapters.

Since the surface concentration, C_{OX}^s, is zero while the solution being swept toward the surface is C_{OX}^*, a finite value, it follows that the final step in the transport process must still be diffusion, controlled by this concentration difference acting over a very small distance, the *diffusion layer*, δ. The current can be expressed in terms of δ by:

$$I(t) = \frac{nFAD_{OX}}{\delta} \cdot C_{OX}^*(t) \tag{10-4}$$

It is instructive to equate the two expressions for $I(t)$ in Eqs. (10-3) and (10-4), which gives us:

$$\frac{dC_{OX}^*}{dt} = -\frac{AD_{OX}C_{OX}^*}{V\delta} = -\frac{J}{V} C_{OX}^* \tag{10-5}$$

where $J = AD_{ox}/\delta$, a constant of the experiment. Rearrangement and integration give:

$$C_{OX}^*(t) = C_{OX}^*(0) \exp\left(-\frac{J}{V}t\right) \tag{10-6}$$

This expression describes mathematically a process of depletion as electrolysis continues. It is characterized by a half-life which is given by:

$$t_{1/2} = 0.693 \,\frac{V}{J} \tag{10-7}$$

After this length of time, the concentration is reduced by one half.

Equation (10-6) can also be used to calculate the current by substituting into Eq. (10-4) to give:

$$I(t) = \frac{nFAD_{OX}}{\delta} \cdot C_{OX}^*(0) \exp\left(-\frac{J}{V}t\right) \tag{10-8}$$

or

$$\ln I(t) = \ln\left(\frac{nFAD_{OX}}{\delta} \cdot C_{OX}^*\right) - \frac{J}{V}t = \ln\left(nFJC_{OX}^*\right) - \frac{J}{V}t \tag{10-9}$$

But we know from Eq. (10-4) that, at $t = 0$,

$$I(0) = nFJC_{OX}^*(0) \tag{10-10}$$

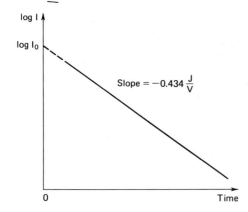

Figure 10-1. Current–time relations for constant-potential electrolysis.

So we can write:

$$I(t) = I(0) \exp\left(-\frac{J}{V}\, t\right) \tag{10-11}$$

from which

$$\log I(t) = \log I(0) - 0.434\,\frac{J}{V}\, t \tag{10-12}$$

Hence, a plot of $\log I$ against t will give a straight line of slope $-0.434\,J/V$, with an intercept at $t = 0$ equal to $\log I(0)$ (Figure 10-1). This enables one to determine $t_{1/2}$ once the slope is known, and thus the duration of the electrolysis needed to diminish the concentration to some desired fraction. These considerations apply only where the product of the reaction is soluble.

Electrolysis can be used for a variety of preparative purposes, but the equations must be altered to reflect the production rather than the disappearance of reagent. It provides a convenient procedure for the preparation of unstable reagents *in situ*.

COULOMETRY

In accordance with the laws of electrochemical equivalence stated by Faraday in 1833, the amount of chemical effect caused by an electric current passing through an ionic liquid is proportional to the quantity of charge, Q, passed. Since the charge (in coulombs) is given by the product of the current (in amperes) and the time (in seconds), we can write:

$$\text{Number of moles reacted} = \frac{Q}{nF} = \frac{1}{nF}\int_0^t I(t)\,dt \tag{10-13}$$

The factor $1/nF$ appears here simply as a proportionality constant, but this is, in fact, the defining relation for the familiar Faraday constant.†

Let us consider the familiar reaction $OX + ne^- \longrightarrow RED$. If we assume no RED to be present initially, the concentration at time t can be obtained by dividing both sides of Eq. (10-13 by V, the volume of the solution, to give:

$$C_{RED}^*(t) = \frac{1}{nFV}\int_0^t I(t)\,dt \tag{10-14}$$

This relation provides an absolute method of making a quantative chemical determination, in the sense that the measurement of the current-time integral can readily be carried out with great precision without the need of a chemical standard of comparison.

It is essential in coulometric analysis that the fraction of the current producing the desired chemical effect be accurately known. Ordinarily, this requires that the fraction be 100 percent, implying that *no* side reactions occur. If this is true, the system is said to be *100 percent current-efficient*.

Three cases can be considered for coulometric analysis: (1) the current is controlled (usually held constant); (2) the potential of the working electrode is controlled; and (3) neither is closely controlled. Case (3) is seldom employed and will not be treated further.

Controlled-Potential Coulometry

In this method, a potential is selected by the same criteria mentioned previously in connection with electrodeposition, and the current-time curve developed will be that of Figure 10-1.

One can obtain from Eq. (10-11) by integration:

$$Q = I(0) \int_0^t \exp\left(-\frac{J}{V}t\right) dt = \frac{I(0)}{J/V}\left[1 - \exp\left(-\frac{J}{V}t\right)\right] \tag{10-15}$$

A plot of Q as a function of time (Figure 10-2) shows that the quantity of charge consumed approaches asymptotically the limiting value of $VI(0)/J$, as t increases. By Eq. (10-10) this is seen to be equal to $nFVC_{OX}^*(0)$ from which the concentration can be obtained. To make such a plot, we need a continuously operating coulometer with its output fed to a strip-chart recorder.

The controlled potential coulometric technique is widely applicable to the determination of small concentrations of electroactive species where separation by potential is advantageous. An excellent example [2] is the determination of plutonium (6–12 mg aliquots) to a relative standard deviation of 0.06 percent. An ion-exchange procedure was needed to remove interferences. The referenced paper gives details of a data treatment method that is capable of much higher precision than the graphical procedure.

†For an account of a modern, highly precise, determination of the Faraday constant, see the paper by Bower and Davis, of the National Bureau of Standards [1].

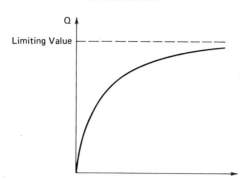

Figure 10-2. A plot of coulometer output, Q, against time in a constant-potential coulometric analysis.

Constant-Current Coulometric Titration

Constant-current operation is primarily useful for coulometric titrimetry, in which the conventional volumetric addition of a reagent is replaced by electrochemical generation.

Coulometry at constant current offers a major advantage in its instrumental simplicity. The current–time integral, which measures the amount of reactant, is merely the product of current and time, so the coulometer can consist of a simple timer. Some provision must be made to prevent the cathode (in a reduction) from becoming so negative that current efficiency is impaired. This goal can be achieved through the addition of an excess of a second substance that will be reduced at a potential slightly more negative than the substance being sought, to act as an intermediate. An example will make this clear.

Suppose that we wish to determine cerium by reduction of Ce(IV). A voltammetric scan with platinum electrodes, with stirring, gives a current–voltage plot such as curve 1 in Figure 10-3, in which the height of the plateau is proportional to the concentration of Ce(IV). If current at level a is passed through the solution, Ce will be reduced initially at about +1.2 V (point A). As the electrolysis proceeds, the concentration of Ce(IV) diminishes until the plateau of the curve drops below a (curve 2). The potential now shifts to about -0.3 V (point B), where H^+ is reduced together with the cerium, and the current efficiency drops below 100 percent.

Let us repeat the experiment with the addition of an excess of ferric iron. This species alone would give a voltammogram like curve 3. Now, when the current causes the Ce(IV) plateau to drop below a, the potential will increase only to about +0.44 V (point C), and Fe(III) begins to be reduced to Fe(II), while cerium continues to be reduced from IV to III. Full current efficiency is maintained because no H_2 is evolved and the secondary reaction:

$$Ce(IV) + Fe(II) \longrightarrow Ce(III) + Fe(III) \tag{10-17}$$

takes place quantitatively, so that each electron passing through the electrode corresponds to one ion of cerium reduced.

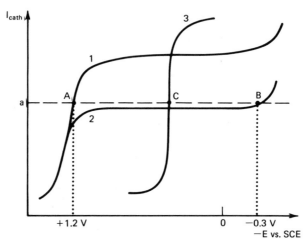

Figure 10-3.   Voltammetric curves of (1 and 2) Ce(IV), (3) Fe(III), in $1 M H_2SO_4$, at a platinum cathode.

As in any form of titrimetry, means must be at hand for identifying the equivalence point. In general, the requirements are no different for coulometric titrations than for their volumetric counterparts. Any of the usual endpoint detection techniques can be used. The accuracy by which the observed endpoint agrees with the theoretical equivalence point must be proven, just as in any other kind of titration.

Instrumentation

The fundamental instruments for electrodeposition are the three-electrode potentiostat for controlled-potential methods (Figure 10-4) and the corresponding galvanostat where the current is to be controlled (Figure 10-5). The only feature in which they are likely to differ from the prototypes described in Chapter 1 is in

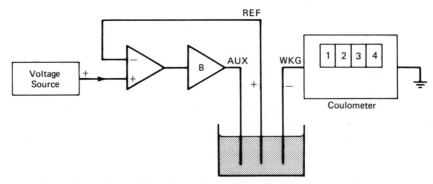

Figure 10-4.   Potentiostat for coulometric analysis. For a gravimetric determination, the coulometer could be omitted and the working electrode grounded directly. Amplifier "B" is a booster.

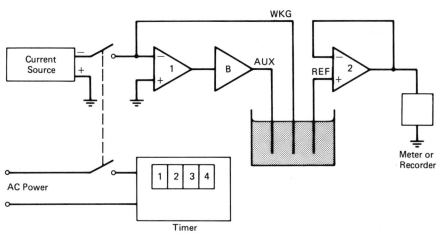

Figure 10-5. Galvanostat for coulometric titrimetry. If a time-based recorder is used, the separate timer can be omitted.

the need for a *booster* amplifier to permit handling greater currents than the usual operational amplifier can accommodate. Conceptually, the booster can be thought of as merely a power-handling output stage of the control amplifier.

The working electrode must be large enough to handle the expected current without excessive current density that would increase the overvoltage and hence result in poorer resolution. Recall that in the expressions for charge-transfer kinetics it is the current density, I/A, rather than the current *per se* that determines the overvoltage η_t. Effective stirring is essential and should be maintained at a constant and reproducible rate.

For coulometric titration, the current is usually not in excess of 15 or 20 mA, and small platinum foil electrodes will suffice. The counter electrode frequently must be isolated from the main solution, so that the product of oxidation will not be stirred into the neighborhood of the working electrode where it might invalidate the analysis. Such isolation often can be accomplished effectively enough by means of a length of glass tubing with a fine-porosity frit at its end. This is particularly satisfactory for oxidative titrations in a buffered medium, since the hydrogen gas produced at the cathode escapes readily and the OH^- ions also produced are absorbed by the buffer. Diffusion of an undesirable ion through the frit can sometimes be eliminated by the judicious use of a layer of ion-exchange resin on the frit. For titrations where the highest precision is required, separate anode and cathode compartments must be provided, connected by one or even two flushable salt bridges to avoid cross-contamination completely [3].

Coulometers. In principle, any type of electrical integrator can be adapted as a coulometer. The most versatile is based on an operational amplifier with capacitive feedback (Figure 10-6). The amplifier forces the incoming current to charge the capacitor in accordance with the formula:

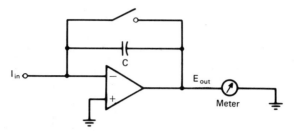

Figure 10-6.   A simple op-amp integrator.

$$E_{out} = -\frac{1}{C} \int I_{in} \, dt \qquad (10\text{-}18)$$

E_{out} can be connected to a digital panel meter or a recorder, and will read out directly in coulombs. This simple integrator is usable only with small quantities of charge, since beyond this the required capacitor becomes too large for convenience.

For larger currents, the circuit of Figure 10-7 can be used instead. The input amplifier (#1) integrates the incoming current, just as in the previous circuit, but when the voltage on the capacitor reaches 10 V, amplifier #2 will abruptly shift its output from positive to negative, actuating the relay coil. This does two things: it increments the counter, and simultaneously discharges the capacitor, resetting it to zero, ready for the next charge. If the charge on the capacitor at 10 V is q_c, then the total number of coulombs is given by the reading on the coulometer multiplied by q_c.

A coulometer working on a similar principle is shown in Figure 10-8. The incoming current is converted by the amplifier to an equivalent voltage, and this is fed to a voltage-to-frequency (V/F) converter. The V/F unit produces a train of pulses at a frequency that is proportional to its input voltage and hence to the current to be evaluated. It is only necessary, then, to count the number of pulses during an experiment to have a measure of the coulombs passed.

Any of these circuits can accept a varying current and give an output reading

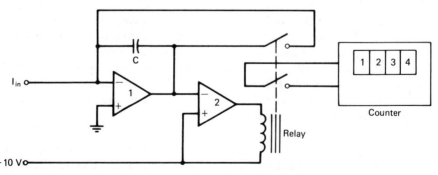

Figure 10-7.   An op-amp integrator with the dynamic range increased by the use of a counter. The relay must be of the polarized type.

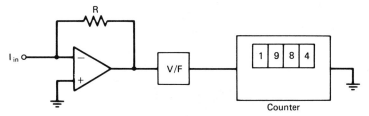

Figure 10-8. A coulometer based on frequency conversion.

proportional to its time-integral. For a constant current, the timer shown in Figure 10-5 will suffice, since the number of equivalents involved is directly proportional to the time.

REFERENCES

1. V. E. Bower and R. S. Davis, *J. Res. Nat. Bur. Standards*, **1980**, *85*, 175.
2. M. K. Holland, J. R. Weiss, and C. E. Pietri, *Anal. Chem.*, **1978**, *50*, 236.
3. G. Marinenko and J. K. Taylor, *Anal. Chem.*, **1968**, *40*, 1645.

Chapter 11

METHODS WITH CONVECTION: II. HYDRODYNAMIC VOLTAMMETRY

As we have seen, for voltammetric experiments in quiet solutions the current response is a transient that decays with time. As the current approaches zero, the information available disappears. In order to obtain more information, the redox process must be forced to continue. This can be accomplished by changing the voltage, as is done in LSV. Another possible approach is to provide fresh solution to the electrode by mechanical means. The term *hydrodynamic* is used to describe this procedure, since solution flow is invariably associated with it.

Methods using convection have already been examined in the previous chapter, and now we shall concentrate on those cases in which only small currents are passed, and the electrode only serves as a measurement probe.

CLASSIFICATION OF METHODS

A possible scheme for the classification of mechanisms for bringing fresh material to the surface is the following:

Interface Renewal

One such procedure employs intermittent stirring of the solution, with application of the voltage during quiet periods. A plot of the current against voltage can be used in a similar manner to a polarogram [1]. This has the disadvantage that even though the solution is renewed, the electrode surface is not changed, and any contaminants will accumulate on it. This can be alleviated by scraping the electrode surface [2], which also contributes some stirring of the solution. Another complicated procedure has been reported [3] that made use of a revolving drum bearing

24 electrodes to be dipped sequentially into the solution. However, the only really successful continuously renewed electrode is the DME (and the SMDE), which provides a sequence of electrodes, the individual drops, that form automatically at the tip of the capillary.

Flow-Through Systems

In a second type of experiment, the electrolyte is actually enclosed within a tubular electrode [4], a porous electrode [5, 6], or one formed from a conductive fabric [7]. The latter two can be considered to be collections of tiny tubular electrodes with complex paths for the solution to take. The displacement of the solution provides the desired replenishment of the active species. Flow-through cells are useful for preparative purposes and also for detectors in chromatography and process control [8]. They are inherently convenient to automate in flow systems.

Continuously Stirred Systems

A very useful arrangement is obtained when the bulk of the solution moves continually past the surface, as with a rotating electrode. This differs from the previous case in that the volume of the liquid involved is usually much larger and the electrode interacts repeatedly with the whole solution. The stirring can be implemented in a variety of ways using vibration or rotation. The DME can be submitted to either form of motion [9, 10]. The most popular configuration for solid electrodes is the rotating disk electrode (RDE). It has the advantage of relative simplicity, commercial availability, and ease of theoretical treatment.

DISK ELECTRODES

A Basic Experiment

The most common form of RDE is a rotating platinum disk, as shown in Figure 11-1. The motion of the disk causes the liquid to move the same way, thus ensuring a continuous flux of matter to the electrode. For any combination of potential, speed, and concentration, there will be a corresponding steady-state current, in contrast to unstirred systems where the current drops off with time. Scanning the voltage results in a current that follows a step-shaped curve, the height of which is proportional to the concentration of the active species. This method has a scope similar to that of polarography, and can be used for any cases where a solid electrode is desirable.

In its motion the disk drags with it a layer of liquid of effective thickness P, called the *fluid boundary layer*. The motion of this layer of liquid is complicated by the fact that centrifugal forces move the liquid away, while fresh solution continually replaces it. As a result, a given molecule first rises into the fluid boundary layer, reacts, and then proceeds to the perimeter of the disk, following a spiral

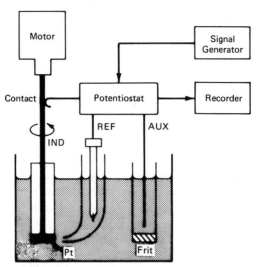

Figure 11-1. A rotating disk electrode system. The Luggin capillary and the fritted glass isolator can often be omitted.

path. The thickness P is of the order of $3(\nu/\omega)^{1/2}$, where ν is the *kinematic viscosity* (defined as the viscosity divided by the density), while ω is the angular velocity [11].† As an example, if $\nu = 0.01$ cm^2 s^{-1} and $\omega = 10$ rad · s^{-1}, the value of P comes out to be about 1 mm. Since the cell dimensions are usually much larger than this, the walls may be considered to be at infinity.

Within the boundary layer the concentration is essentially equal to the bulk concentration except for a thin portion in the immediate vicinity of the electrode, where its variation between the bulk value, C^*, and the surface value, C^s, is nearly linear. This is illustrated in the calculated profile [11] shown in Figure 11-2. If one approximates this concentration profile by two straight lines, their intersection will define the thickness of the *diffusion layer* or *Nernst layer*, δ. Theory shows this to be given approximately by:

$$\delta \cong 1.6(D/\nu)^{1/3} (\nu/\omega)^{1/2} \qquad (11\text{-}1)$$

For example, if $D = 10^{-5}$ cm^2 s^{-1} and $\omega = 10$ rad s^{-1}, then the thickness of the diffusion layer will be about 0.05 mm, or $\frac{1}{20}$ of the value calculated above for P. This fraction does not change much from experiment to experiment, since both D and ν tend to remain constant. This formula indicates also that the thickness of the Nernst layer varies, following an $\omega^{-1/2}$ dependence, and becomes semi-infinite in the absence of rotation ($\omega = 0$). The transport at distances less than δ is primarily only by diffusion, while beyond δ convection predominates (Figure 11-3). In both cases, the process rapidly attains a steady state. This contrasts with polarography in its various forms, where transients are fundamentally involved in the

†Some authors use the numerical constant 3.6 rather than 3.

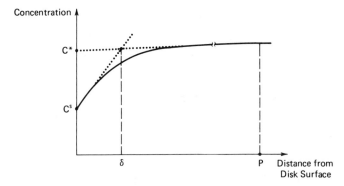

Figure 11-2. Concentration profile at a rotating disk electrode. The distance δ is much smaller than P.

measurement operation. The stability of the output in polarography is not due to stationary states, but to readings taken at precisely the same phase in the life of successive transients.

The above discussion implies *laminar flow*, in which the liquid moves in parallel layers. In reality, there will always be a certain amount of turbulence caused by surface roughness, gas bubbles, and so on, but such local perturbations have short lives and tend to dissipate, so that laminar flow is quickly reestablished. As the rotation rate increases, the turbulence becomes more and more long-lived, until it becomes self-sustaining. The degree of stability toward turbulence is described by the Reynolds number, Re, given for the rotating disk by $Re = r^2(\omega/\nu)$ where r is the distance from the axis of rotation. In the example above, if $r = 1$ cm and $\omega/\nu = 10^3$, the value of Re becomes 1000. Such a small value indicates laminar flow. For $Re > 10\,000$, the flow is partly turbulent, and at values beyond about 5×10^5 it is completely turbulent. Most electrochemical work is carried out under laminar flow conditions, but sometimes turbulence is desirable because the electrical current can be much larger.

Theoretical Considerations

For a steady-state current to be established in a stirred solution, one must scan slowly or pause long enough after each step for any transient to decay. The situation

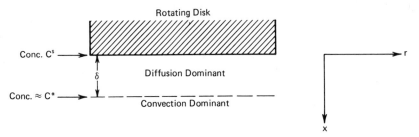

Figure 11-3. Model for convective diffusion.

is complicated by the fact that both hydrodynamic and chemical steady states must exist simultaneously. The major factors to be considered are:

1. The time needed to establish the flow patterns. This is surprisingly small, since the trajectories of the fluid molecules become well established within a single turn of the disk, typically in tens of milliseconds.

2. The time required for the formation of a new concentration profile following a change in potential or in the rate of rotation. The duration of this type of transient [12] is of the order δ^2/D. If the Nernst layer is 0.005 cm thick and D is 10^{-5} cm^2 s^{-1}, the time will be 2.5 seconds. This determines in many cases the time needed for the faradaic current to reach its stationary state.

3. The charging current needed to modify the double-layer charge to a new value as required by changes in voltage. In most experimental systems, this is a short transient with time constants on the order of milliseconds or even less. Consequently, charging currents play little part in hydrodynamic systems.

4. Alteration of surface conditions. Solid electrodes often undergo surface changes, such as oxidations, leading to perhaps a monolayer of products. Adsorption from the solution can also contribute to current transients. In either case, the equilibrium is approached rather slowly, in seconds or even minutes. This type of transient appears when the voltage is changed but not when the rotation speed alone is altered.

The protracted nature of many of the transients listed above limit the rate at which measurements can be carried out. The duration of an experiment must usually be some 50 to 100 times the total relaxation time to ensure that a steady state is continuously maintained as the voltage is scanned.

Once a steady state is attained, kinetic control may belong to the charge-transfer reaction. In this case,

$$I = nFAk_fC^s \tag{11-2}$$

where k_f is the forward rate constant including the exponential voltage-dependent factor. This type of control occurs for small overvoltages and for small rate constants. When the rate constant is larger (the redox process is rapid), control shifts to the so-called *convective diffusion* in which both convection and diffusion play a part, the slower of these being diffusion across the Nernst layer.

Mass transfer can be estimated by means of the simplified model shown in Figure 11-3, which assumes the concentration profile described by the dotted line of Figure 11-2. The surface concentration is taken to be C^s across the whole rotating disk electrode, and the concentration profile is assumed to be independent of the position on the disk.

In this case Fick's law can be written (in one dimension) as:

$$\text{Flux} = \Phi = D\frac{dC}{dx} \tag{11-3}$$

where x is the distance from the disk. The flux can be expressed in terms of the current by the relation $\Phi = I/nFA$, so that:

$$\frac{I}{nFA} = D\frac{dC}{dx} \tag{11-4}$$

Integration between zero and the Nernst layer thickness δ gives:

$$\frac{I\delta}{nFA} = D\,(C^* - C^s) \tag{11-5}$$

or

$$I = nFA\,(D/\delta)\,(C^* - C^s) \tag{11-6}$$

Use can be made of the previously introduced expression for:

$$\delta = 1.6\,(D/\nu)^{1/3}\,(\nu/\omega)^{1/2} \tag{11-7}$$

which gives us the equation:

$$I = 0.62\,nFAD^{2/3}\nu^{-1/6}\omega^{1/2}(C^* - C^s) \tag{11-8}$$

where the concentrations refer to either OX or RED as appropriate. This equation was first derived by Levich using a rigorous mathematical treatment of the transport equations.

The voltage dependence of the current is reflected in a curve quite similar to the envelope of a polarogram and can be characterized by an $E_{1/2}$. The height of this wave is the *limiting current*, i_d, equivalent to the diffusion current in polarography:

$$i_d = 0.62nFAD^{2/3}\nu^{-1/6}\omega^{1/2}C^* \tag{11-9}$$

FLOW-THROUGH ELECTRODES

The current drawn by a tubular electrode such as that illustrated in Figure 11-4a, has a form similar to Eq. (11-9), except that instead of ω, the experimentally controllable variable is now the volume of liquid passed per unit time, the *volumetric flow*, V_f. The pertinent relation is:

$$i_d = 5.2 \times 10^5\,nD^{2/3}X^{2/3}V_f^{1/3}C^* \tag{11-10}$$

where X is the length of the tube. It is interesting to note that the tube diameter does not appear explicitly. This is because an increase in the diameter is accompanied by a reduction of the fluid velocity for a given V_f, and the two effects com-

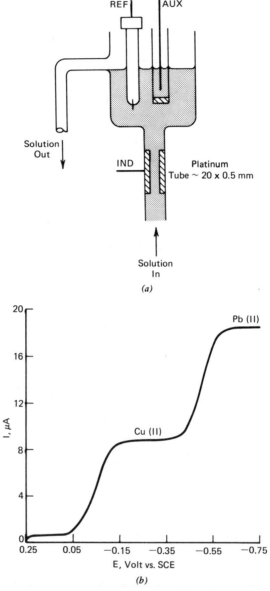

Figure 11-4.   (a) A possible arrangement of a tubular electrode. (b) Voltammetric curve of a solution 5×10^{-5} M each in Cu(II) and Pb(II) in 0.1 M KNO$_3$, at a mercury-plated platinum electrode.

pensate each other.   The current increases with the flow rate, but if laminar flow is to be maintained, V_f must be kept rather small.   For an electrode such as that illustrated in Figure 11-4, a reasonable flow rate would be about 0.2 cm^{3}s^{-1}.   Even with this small flow, a solution 10^{-6} M in the active species would give a current of perhaps 0.25 μA, considerably larger than the corresponding DC polarographic diffusion current. An example [13] of voltammograms obtained with a mercury-covered tubular platinum electrode is shown in Figure 11-4b. It is evident that concentrations in the micromolar range could easily be measured.

Porous electrodes are similar to their tubular counterparts. They can be constructed of a fine metallic mesh or of a column filled with conductive granules. Alternatively, one can use a cylinder made of reticulated vitreous carbon (a porous form of graphite) clad in an impermeable sheath. A mesh or a short column will exhibit the same flow dependence, $V_f^{1/3}$, as the tube [Eq. (11-10)]. This is understandable, since a porous electrode can be thought of as consisting of a collection of tubes acting in parallel. For longer columns, the reaction eventually approaches completion, and the current becomes proportional to the amount of active species within the pores and, in turn, to the first power of the flow rate:

$$I = nFV_f C^* \qquad (11\text{-}11)$$

Experimental Techniques

The only significant differences between DC voltammetry in quiet and in stirred solution, in terms of instrumentation, lie in the cell system. A DC polarograph could be used for either purposes, as seen in Figure 11-1.

In contrast, if modulation is required, the instrumentation requirements are different. In hydrodynamic voltammetry there are two ways in which modulation can be acheived: (1) *Hydrodynamic modulation (Nernst layer modulation)*, where the diffusion layer is altered periodically by changing the rate of rotation, ω, or the fluid flow rate, V_f. The resulting changes in δ are reflected in changes in the concentration slope, dC/dx, that in turn are followed by the current, since $I = nFA(dC/dx)$. (2) *Voltage modulation (Nernst equation modulation)*, the conventional method in which the applied voltage is varied according to a predetermined program. By the Nernst equation, the ratio of the concentrations of OX and RED will follow this variation, thus changing the slope of the concentration profile.

Voltage modulation is rather unsuccessful because of the long transients encountered, and is not often used. In contrast, hydrodynamic modulation is quite effective. Changes in ω and in V_f produce transients lasting, at the most, a few seconds, so that modulation in the frequency region of 0.1 to 1 Hz can be used without deviating from steady-state conditions.†

We shall restrict our discussion to flow modulation. Consider Eq. (11-8), and assume that the angular velocity is modulated with a small amplitude in such a way that a sinusoidal current response is generated. This parallels AC polarography. To generate the sine wave, it is convenient to use a function of type:

$$\omega = [A + B \sin (2\pi f_{mod} t)]^2 \qquad (11\text{-}12)$$

---

†There is no compelling reason for the use of steady-state operation, except for the elegance of the mathematical treatment. On the other hand, the response decreases beyond a characteristic frequency of modulation, much like the attenuation due to an electrical low-pass filter.

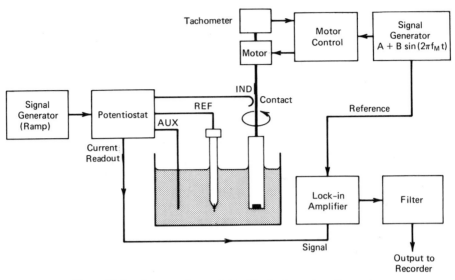

Figure 11-5. A system for modulated hydrodynamic voltammetry.

Where A and B are constants and f_{mod} is the frequency of the modulation. The square-root that appears in the Levich equation then gives:

$$\omega^{1/2} = A + B \sin{(2\pi f_{mod} t)} \tag{11-13}$$

leading to a current equation of the form:

$$I = (\text{const})\,[A + B \sin{(2\pi f_{mod})}]\,(C^* - C^s) \tag{11-14}$$

Consequently, the current has two components, the original DC determined by A, and a superimposed AC defined by the term in B, both dependent on the concentration. It follows that the AC component is not a derivative quantity as in AC polarography but gives the same type of wave as does DC.

The rate of rotation, ω, is commonly given in RPM rather than radians per second, while f_{mod} is in Hertz. Consider an example [15]: If the value of A is 60 $(\text{RPM})^{1/2}$ and B is 6 $(\text{RPM})^{1/2}$, and if f_{mod} is 3 Hz, then the rotation of the disk will vary sinusoidally between 3564 and 3536 RPM at a repetition rate of 3Hz.†

The instrument in Figure 11-5 uses a lock-in amplifier, which is a kind of super filter. The output of the amplifier contains only the components of the current that are at the frequency f_{mod} and that have the desired phase relation compared to the reference.

A resulting graph [15] is shown in Figure 11-6, compared with the results of the unmodulated case. It is evident that the hydrodynamically modulated RDE is effective in diminishing the slope of the baseline. From the rather clean wave ob-

†This is an example of frequency modulation.

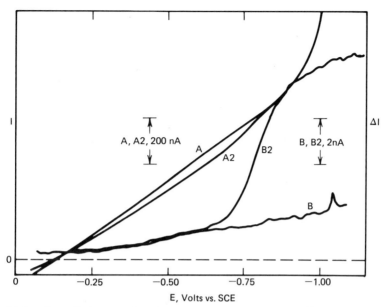

Figure 11-6. Controlled potential cathodic scans of Tl(I) at a rotated amalgamated gold disk electrode. Curves A and A_2 without modulation, B and B_2 with hydrodynamic modulation. Curves A and B are blanks, curves A_2 and B_2 are for $2.0 \times 10^{-7} M$ Tl(I) in $0.1 M$ HClO$_4$. (Analytical Chemistry [15].)

tained with the $2 \times 10^{-7} M$ solution, one can conclude that as low as $10^{-8} M$ would be attainable. The AC signal is smaller than the DC by the factor B/A, as can be seen from Eq. (11-14); B/A is typically about 0.1. This loss is more than compensated for by the elimination of noise and charging current contributions.

Again paralleling polarography, it is possible to devise a system for square-wave hydrodynamic modulation. This can be applied to tubular electrodes as well as to the RDE; it consists in abruptly changing either ω or V_f between two limits [16]. The resulting sensitivity is similar to that obtained with sine-wave modulation, about $10^{-8} M$. The current response is not a derivative, and waves are obtained as output.

Applications

Hydrodynamic voltammetry can be used for direct analysis of reducible or oxidizable species, but its merits are more evident in two applications: electrode kinetics and the analysis of flowing solutions. We shall discuss them in turn.

Electrode Kinetics.   Kinetic and mechanistic studies using rotating disks frequently employ an alternative construction, wherein a ring-shaped electrode surrounds the disk, as shown in Figure 11-7. Such a double electrode gives an added dimension to the experiment, since the process can now be analyzed simultaneously by two measurements: the disk current to give the overall process, and the ring current

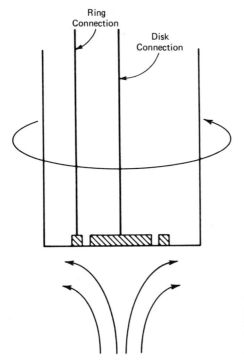

Figure 11-7. A ring-disk electrode. The cross-hatched parts are the electrode itself (Pt), set in an insulating material such as Teflon. The electrical contacts must allow for the rotation of the electrode.

through which the products of the disk reaction can be estimated. The ring can be set at any desired potential, it can be made, for instance, to reoxidize the product of reduction generated at the disk. This is somewhat akin to cyclic voltammetry, in which the potential is reversed to permit analysis of the products of the forward scan. The difference is that the time lapse is much shorter in the ring-disk system, since the transit time of the liquid across the gap separating them is very short (milliseconds).

The experiment requires a special type of potentiostat to accomodate four electrodes: a reference, an auxiliary, and two working electrodes that can operate at two different potentials, E_r and E_d, (Figure 11-8). A good strategy is to maintain one of the working electrodes at instrumental ground and the other at an off-set potential. This circuit gives a partial IR-compensation.

The ring current is given [17] by an equation similar to that for disks alone, also derived by Levich:

$$I = 0.62 \, nF\pi(r_3^3 - r_2^3)^{2/3} D^{2/3} \nu^{-1/6} \omega^{1/2} (C^* - C^s) \qquad (11\text{-}15)$$

where r_2 and r_3 are the inner and outer radii of the ring, respectively. This should be compared to Eq. (11-8). If opposite reactions occur at ring and disk, even though the two currents will not as a rule be equal, there will be a constant ratio

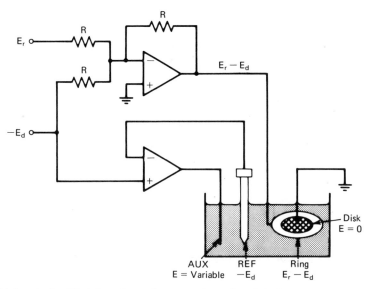

Figure 11-8. A simplified four-electrode system for a ring–disk electrode. The potential of the disk (vs. SCE) is $0 - (-E_d) = E_d$, while the ring potential is $(E_r - E_d) - (-E_d) = E_r$. Not shown are the additional circuits needed to measure the individual currents of the ring and the disk.

between them that permits exact determination of the results. This is called the *collection efficiency*, U:

$$U = I_{ring}/I_{disk} \qquad (11\text{-}16)$$

The ring current is free from charging components and can serve to identify the products of the disk reaction with sufficient sensitivity to permit the study of monolayer reactions [18] or adsorption–desorption effects. The ring–disk pair is quite versatile for kinetic studies, since there are three controllable variables, E_r, E_d, and ω, in addition to the solution variables. There are also two outputs, I_r, and I_d, that can be used to establish the mechanism of a reaction.

The experiment is sometimes carried out with $E_r = E_d$, in which case the two processes are in competition, a so-called *shielding* experiment. If the rotation rate is modulated, the improvement in baseline is comparable to experiments with the disk alone.

Flow-Detectors. For flow-lines, such as the output from an HPLC or a process control system, electrochemical detectors possess advantages of sensitivity and simplicity over many alternative devices. Detectors in such applications must have a small internal volume, since the time needed to replace the content of the detector with fresh sample determines the limit of resolution for flow systems. For example, in monitoring chromatographic peaks, if the internal volume is V_{det}, the volume corresponding to a peak on the chromatogram must be larger than V_{det} by a factor

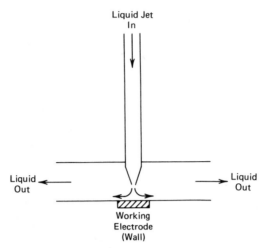

Figure 11-9. Example of a wall-jet electrode. The counter and reference electrodes can be connected at the liquid outlet.

of 10 or so. In practice, the volume can easily be made as small as 10 μL, so that peaks of 100 μL can be satisfactorily measured.

The applicability of tubular electrodes to organic species is broad. It appears that reasonable discrimination is available to obtain response from, e.g., a nitro compound in the presence of ketones. Although DME-based flow cells have been used successfully, hydrodynamic cells are inherently more adaptable to such applications, especially the tubular electrode mentioned above, and the *wall-jet* electrode [19]. This latter device is shown in Figure 11-9. The very fast flow seems to prevent contamination of the platinum target. The dead volume can be as little as one microliter.

REFERENCES

1. S. Roffia and E. Vianello, *J. Electroanal. Chem.*, **1966**, *12*, 112.

2. N. D. Tomashov and L. P. Vershinina, *Electrochim. Acta*, **1970**, *15*, 501.

3. R. J. Lawrence and J. Q. Chambers, *Anal. Chem.*, **1967**, *39*, 134.

4. W. J. Blaedel, C. L. Olson, and L. R. Sharma, *Anal. Chem.*, **1963**, *35*, 2100.

5. R. E. Sioda, *Electrochim. Acta*, **1970**, *15*, 783.

6. A. N. Strohl and D. J. Curran, *Anal. Chem.*, **1979**, *51*, 353.

7. D. Yaniv and M. Ariel, *J. Electroanal. Chem.*, **1977**, *79*, 159.

8. R. J. Rucki, *Talanta*, **1980**, *27*, 147.

9. Y. Okinaka and I. M. Kolthoff, *J. Am. Chem. Soc.*, **1957**, *79*, 3326.

10. H. J. Mortko and R. E. Cover, *Anal. Chem.*, **1979**, *51*, 1144.

11. A. C. Riddiford, in "Advances in Electrochemistry and Electrochemical Engineering" (P. Delahay, ed.), Wiley-Interscience, New York, **1966** Vol. 4, p. 47.

12. S. Bruckenstein and S. Prager, *Anal. Chem.*, **1967**, *39*, 1161.
13. T. O. Oesterling and C. L. Olson, *Anal. Chem.*, **1967**, *39*, 1543.
14. K. Tokuda and S. Bruckenstein, *J. Electrochem. Soc.*, **1979**, *126*, 431.
15. B. Miller and S. Bruckenstein, *Anal. Chem.*, **1974**, *46*, 2026.
16. W. J. Blaedel and R. C. Engstrom, *Anal. Chem.*, **1978**, *50*, 476.
17. S. Bruckenstein and B. Miller, *Acc. Chem. Res.*, **1977**, *10*, 54.
18. B. Fleet and C. J. Little, *J. Chromatog. Sci.*, **1974**, *12*, 747.
19. J. Yamada and H. Matsuda, *J. Electroanal. Chem.*, **1973**, *44*, 189.

Chapter 12

STRIPPING ANALYSIS

In the analysis of very dilute samples, it is often necessary to employ some type of preconcentration step prior to actual analysis, whereby the bulk of the diluent is removed. In the context of electrochemistry, this can be accomplished by the electrolytic deposition of trace metals from a large volume of solution onto an electrode, followed by redissolution or *stripping*. Analytical information can be obtained during the stripping process.

THE BASIC EXPERIMENT

Consider an aqueous solution containing trace amounts of many species, among which we are interested in determining the amounts of cadmium and copper. We assume quantities present to be too small (of the order of 10^{-8} M) to determine by direct polarography.

We proceed to plate out the metals by the electrodeposition on a suitable cathode, such as a graphite electrode that has been plated with a thin layer of mercury. The conductivity of the solution must be substantial, so a supporting electrolyte is added. The solution should be sparged with nitrogen. The electrode is connected to a potentiostat at -1.0 V (vs. SCE), whereupon any copper and cadmium present are reduced, forming an amalgam with the mercury.

The electrolysis is allowed to proceed for a precisely known time, say 10 minutes, with good stirring, followed by 30 seconds with the stirring turned off to allow the solution to become quiet. The potential is then scanned anodically while the current is recorded. Figure 12-1 will clarify this sequence of events.

The resulting current–time curve is, in effect, an anodic voltammogram, showing well demarcated peaks corresponding to the reoxidation of successive elements from the amalgam. The outcome of an actual run is shown in a more conventional format for stripping analysis, in Figure 12-2 [1]. Note that the curves for blank and sample are indistinguishable in the region of the Pb peak, showing that the only source of lead is in the reagents themselves. The sample is shown to contain cad-

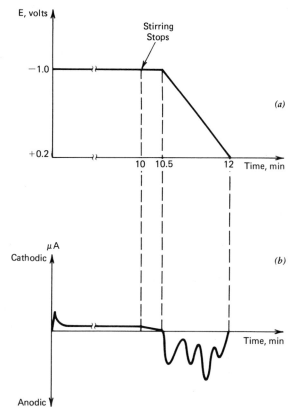

Figure 12-1. (a) Voltage program for anodic stripping voltammetry; the solution is stirred for the first 10 minutes, then quiet. (b) Resulting current, showing anodic peaks corresponding to various electroactive species. The cathodic portion of the current is exaggerated for clarity.

mium and copper but no lead. The method is so highly sensitive that it is essential to run a blank using the same quantities of all reagents and solvents.

THEORETICAL CONSIDERATIONS

Preconcentration (Plating)

The most straightforward way to concentrate the metallic ions from a solution is to employ a potential sufficiently cathodic to cause reduction of all the metals of interest. Current is then allowed to flow for a protracted period, with stirring, to ensure that deposition is quantitatively complete. This constitutes an *exhaustive* procedure, sometimes described as *stoichiometric*. A relatively large cathode is required, since otherwise the electrolysis time will be excessive.

Exhaustive electrolysis is often impracticable, and even when it can be used,

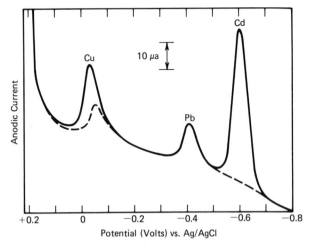

Figure 12-2. Anodic stripping voltammogram. The broken line is a blank, with identical Pb peaks in blank and analytical determination. The solution contained $2 \times 10^{-7}\ M$ Cd^{++} and Cu^{++} in 0.05 M KCl. Electrode: 0.32 cm^2 mercury film. Deposition: 5 min at –1.0 V. Sweep: 2.5 V/min, scanned from right to left. (*Journal of Chemical Education* [1].)

considerable savings in time result from a procedure wherein deposition current is allowed to flow for a specified time interval, rather than until deposition is complete. Hence, the stoichiometric method is seldom employed.

The theory of electrolytic deposition follows the same principles outlined in Chapter 10. It must be realized that for the very small concentrations encountered in the present applications, the Nernst equation predicts potentials much more negative than $E^{\circ\prime}$. For example, a $10^{-7}\ M$ solution of silver in 1 M $HClO_4$ ($E^{\circ\prime} = +0.79$ V vs. SHE) corresponds to a nernstian potential of $0.79 + 0.0592 \log C_{Ag^+} = +0.38$ V.

With a small mercury electrode, such as a hanging drop, the concentration of a reduced metal in the amalgam can become much greater than the concentration of the corresponding ion in the external solution, with the result that the activity coefficients in the two phases may be significantly different. This contrasts with the usual condition in polarography, where these quantities are of the same order of magnitude in the two phases. Hence, allowance would have to be made for this factor if quantitative separations were desired in the deposition step.

If an inert solid, such as platinum, is used as electrode, a phenomenon known as *underpotential* is sometimes encountered. This means that, for very small quantities, a cation is reduced to the metal at a less negative potential than would be predicted by the Nernst equation. This is presumed to occur when the electrode is only partially covered with a monatomic layer of the deposited metal, so that the latter is not in its thermodynamic standard state.

Stripping

The procedure to be followed in reoxidation of deposited metals must meet different requirements for solid electrodes and for mercury. On a solid electrode, in general, since all the metals present have been deposited together, it is likely that some atoms of the more easily oxidizable metals have become buried in a deposit of more noble. In this event, voltage-selective stripping is not practicable. An electrode of this type is best treated by total removal of the deposit by anodic oxidation into a fresh portion of supporting electrolyte, followed by polarographic or other appropriate analytical method.

Reoxidation of metals from an amalgam is relatively straightforward. The process generally used is selective oxidation into the original solution by *reverse-scan voltammetry*. The potential is scanned in the positive direction, starting at the point where deposition took place, as in Figure 12.1.

It is customary to scan at a relatively fast rate, and the Randles–Ševčik equation derived in Chapter 7 for LSV applies to the dissolution of metals from mercury [2]. This equation predicts a current–voltage curve with a series of anodic maxima corresponding to successively oxidized species. The peak potential appears at a point more positive than $E_{1/2}$ by $28.5/n$ mV, and the peak current [3] is:

$$I = 602 n^{3/2} A D_{RED}^{1/2} \, v^{1/2} \, C_{RED}^* \chi_{RED} \tag{12-1}$$

where v is the scan rate, dE/dt, and χ_{rev} is a dimensionless function of potential; at the peak, χ_{rev} takes the value 0.446. Combining numerical constants gives the more specific form of the equation applicable to the peak current (at 25°C):

$$i_{pk} = 269 n^{3/2} A D_{RED}^{1/2} v^{1/2} C_{RED}^* \tag{12-2}$$

(For more details, refer to Chapter 7.)

A correction could be added to Eq. (12-2) to account for spherical rather than planar geometry [3], but the correction is negligible for most purposes (about 2 parts in 10^5 for a typical case).

The significance of this equation is primarily to show the direct proportion between the peak height, i_{pk}, and the concentration of the electroactive species in the amalgam. In this connection, it should be borne in mind that the charging current is *directly* proportional to v, whereas the peak current increases as $\sqrt{v}$. Hence, an increase in the scan rate can be counterproductive if carried too far.

The positive-going ramp voltage in the stripping step can be modulated by pulses, as in polarography [2]. Whereas in the corresponding polarographic method, negative potential pulses are superimposed on the negative-going ramp, in stripping, the working electrode is given a positive-going ramp modulated with positive pulses. Just as in some forms of polarography, the sensitivity is enhanced by the fact that some of the material that is oxidized during a pulse is re-reduced during the subsequent inter-pulse period, since the potential is still cathodic, and hence can be oxidized a second time by the next pulse.

INSTRUMENTATION

The basic electronic elements required for stripping analysis do not differ from the instruments discussed in previous chapters. The only added feature appears in the design of an automated apparatus to accomplish both deposition and reoxidation steps, where precise sequential timing is necessary. Such an apparatus is commercially available. Figure 12-3 shows a possible design.

Electrodes

A variety of electrodes have been found useful in anodic stripping analysis. Because of the restrictions on the use of solid electrodes, such as platinum, which were mentioned above, these electrodes are seldom utilized, and then only for special purposes.

The hanging mercury drop is widely used in spite of the cumbersome and delicate auxiliary equipment needed to form and maintain drops in a reproducible manner. Such an electrode can be fabricated *in situ* by catching a number of droplets from a conventional polarographic capillary in a small scoop (Figure 12-4*a*).

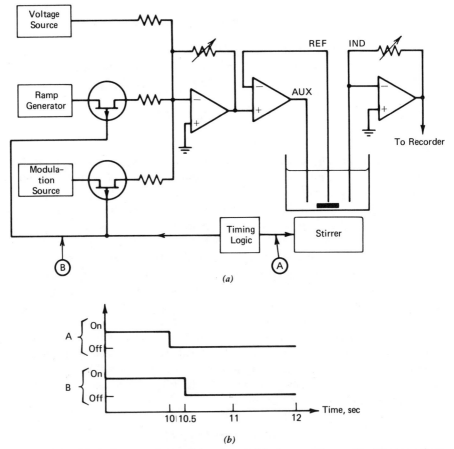

Figure 12-3.   Block diagram of a possible automated instrument for anodic stripping voltammetry. (*a*) Schematic. (*b*) Timing diagram, with specific times corresponding to Figure 12-1.

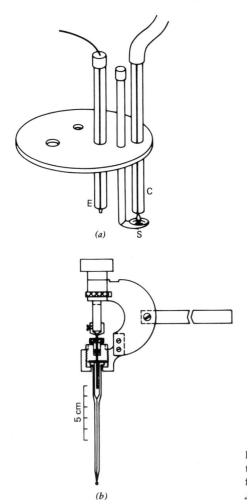

(a)

(b)

Figure 12-4. Typical mercury-drop electrodes for stripping voltammetry. (a) Transfer type. (b) Micrometer type. (b, from
Journal of Electroanalytical Chemistry [5].)

The mercury is transferred manually from the scoop to the tip of a glass-sheathed
platinum wire. A second technique calls for a closed mercury reservoir connected
to a glass capillary and provided with a micrometer-driven plunger (Figure 12-4b)
to force a measured quantity of mercury through the capillary. This method facilitates replacement of one drop by another identical one. The SMDE can be used to
provide hanging mercury drops for stripping voltammetry, simply by resetting a
function switch.

Another type of electrode consists of a thin film of mercury electrodeposited on
a substrate of some form of carbon. An example is the WIG electrode (wax-impreg-
nated graphite). Another is the carbon-paste electrode, made by mixing powdered
graphite with a heavy oil or grease such as Nujol. A third type of carbon electrode
is made from a small piece of so-called *glassy carbon*, epoxied into the end of a
glass tube and polished.

TABLE 12-1
Solubilities
in Mercury [3]
(Atom Percent)

Cd	10.0
Cu	0.006
Ga	3.6
In	70.0
Pb	1.3
Tl	43.0
Zn	5.83

These thin-film electrodes permit shorter plating times because of their large area, but since the volume of mercury is small, there is a real possibility of reaching the solubility limit for a metal being plated. Copper, for example, is only soluble in mercury to the extent of 0.006 atom percent. Data for a few other elements are given in Table 12-1 [3].

A difficulty that may develop with any form of small-volume mercury electrode is the formation of intermetallic compounds between two of the metals that have been deposited on the electrode. Such compounds may not break down readily as the potential becomes more anodic, and will interfere with the normal diffusion process within the mercury. Figure 12-5 shows the kind of anomalous result that may appear [4].

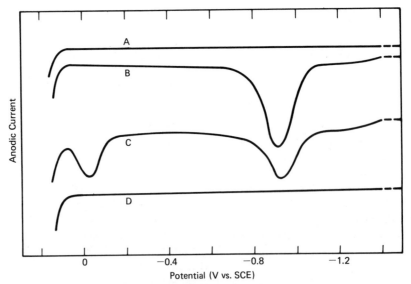

Figure 12-5. Anodic stripping curves for Ni-Zn amalgams. Curve A: supporting electrolyte only (0.1 M KCl). Curve B: 5×10^{-4} M Zn^{++}. Curve C: 5×10^{-4} M each in Zn^{++} and Ni^{++}. Curve D: 5×10^{-4} M Ni^{++}. (*Bull. Acad. Polon.* [4].)

REFERENCES

1. W. D. Ellis, *J. Chem. Educ.*, **1973**, *50*, A131.
2. T. R. Copeland and R. K. Skogerboe, *Anal. Chem.*, **1974**, *46*, 1257A.
3. R. S. Nicholson and I. Shain, *Anal. Chem.*, **1964**, *36*, 706.
4. W. Kemula, Z. Galus, and Z. Kublik, *Bull. Acad. Polon. Sci.*, *Ser. Sci. Chim.*, **1958**, *6*, 661, via I. Shain, "Stripping Analysis," in "Treatise on Analytical Chemistry" (I. M. Kolthoff and P. J. Elving, Eds.), Wiley-Interscience, New York, **1963**, Pt. I, Vol. 4, p. 2544.
5. E. Barendrecht, in "Electroanalytical Chemistry: a Series of Advances" (A. J. Bard, Ed.), Marcel Dekker, New York, **1967**, Vol 2, p. 53.

Chapter 13

CONDUCTOMETRY

Conceptually, the simplest of electroanalytical techniques is the determination of electrolytic conductance,† which provides a measure of the total concentration of ions present. For this reason, it is inherently nonspecific, so that a single measurement of the conductance of a solution containing two electrolytes cannot provide information about either component individually; an additional procedure is required for a complete analysis.

Following its introduction as an analytical tool by Kolthoff in 1923, conductometry became popular as a titration technique, largely as a substitute for the conventional indicator in the determination of acids or bases in the presence of color or turbidity. Subsequently, this use has diminished in favor of more selective methods, and today conductometry is seldom employed in titration. It has, however, found a valuable place in concentration monitoring where selectivity is not required, for example in determining the purity of water. It also provides an important method of detection of effluents in ion-exchange chromatography and related fields.

THE EXPERIMENTAL BASIS

The quantity to be measured in conductometry is a property of the *bulk* of the solution and is not based on an interfacial phenomenon. Hence, there are two possible general approaches to its determination: with or without the aid of electrodes in contact with the solution.

With Electrodes

A typical conductivity cell consists of a pair of platinized‡ platinum electrodes, about 1 cm² in area, mounted in a glass container (Figure 13-1*a*). Each electrode

---

†The conductance is more convenient than its reciprocal, the resistance, since the former is proportional to concentration while the latter is not.
‡The effective surface area of each electrode is increased several hundred times by *platinization*, the electrolytic deposition of a finely divided form of platinum on the surface.

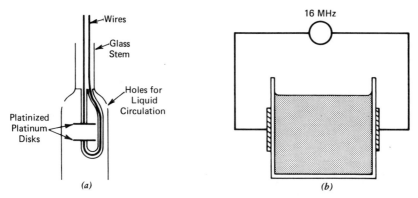

Figure 13-1. Typical conductance cells for analytical applications. (*a*) Dip type. (*b*) High frequency style.

has associated with it an impedance, Z_{far}, shunted by the capacitance of the double layer, C_{dl}, shown in an equivalent electrical circuit in Figure 13-2*a*. The quantity desired to be measured is R_{soln}, the resistance of the solution between the electrodes, which is shunted by the very small interelectrode capacitance C_{soln}. If the source of voltage is DC, then the capacitances can be neglected.

The faradaic impedances, Z_{far-1} and Z_{far-2}, depend on the nature and concentration of the solution, but in general are not negligible. Hence, measurement with DC requires two additional probe electrodes at points X and Y in the circuit, which ensures that only the voltage drop across R_{soln} is measured.

On the other hand, an alternating voltage can be used to advantage. To analyze the circuit, each capacitance must be replaced by its corresponding impedance, $Z_C = 1/(2\pi f C)$. By the law of combination of parallel impedances, we can write for each electrode:

$$Z_e = \frac{Z_{far} Z_{C\text{-}dl}}{Z_{far} + Z_{C\text{-}dl}} \tag{13-1}$$

and for the solution:

$$Z_{soln} = \frac{R_{soln} Z_{C\text{-}soln}}{R_{soln} + Z_{C\text{-}soln}} \tag{13-2}$$

giving rise to the simplified equivalent circuit of Figure 13-2*b*.

Let us now make some assumptions about the numerical values involved. Typical conditions might be as follows:

$$\text{AC amplitude} = 1\text{V}$$

$$f = 1600 \text{ Hz}$$

$$R_{soln} = 100 \ \Omega$$

$$C_{soln} = 10 \text{ pF}$$

$$Z_{C\text{-}soln} = 10 \text{ M}\Omega$$

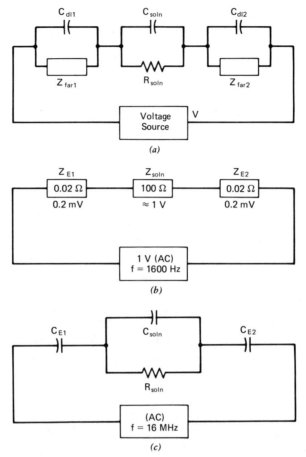

Figure 13-2. (a) The equivalent circuit for a two-electrode conductance cell. (b) Same, for AC, with electrodes (c) Same, for AC, without electrodes.

and for each electrode,

$$Z_{far} = 1000 \ \Omega$$

$$C_{dl} = 5000 \ \mu F$$

$$Z_{C\text{-}dl} = 0.02 \ \Omega$$

From these figures and the equivalent circuit, we can see that essentially the entire 1 volt from the AC source will appear across the solution itself and only a negligible portion of it across each individual electrode interface, namely, about 0.2 mV. This is far too small a voltage to cause chemical effect at the electrodes. The net result is that the ions in solution oscillate back and forth at the frequency of the AC to an extent determined by their relative mobilities, but do not enter into any electrochemical interaction with the electrodes.

Without Electrodes

Since no electron transfer is necessary between electrodes and solution in AC conductance measurements, there is no need for actual contact between the metal surface and the solution. In fact, it is possible to make such measurements with the electrodes replaced by metal plates cemented to the *outside* surfaces of the glass vessel (Figure 13-1 b). In order to keep the capacitive impedance linking the plates with the solution to a reasonable value, a much higher frequency must be employed. The pertinent equivalent circuit is shown in Figure 13-2 c. A set of typical conditions might now be:

$$f = 16\text{MHz}$$

$$R_{\text{soln}} = 100 \ \Omega$$

$$C_{\text{soln}} = 5 \ \text{pF}$$

$$Z_{\text{C-soln}} = 2000 \ \Omega$$

and for each plate (replacing an electrode),

$$C_{\text{plt}} = 15 \ \text{pF}$$

$$Z_{\text{C-plt}} = 6600 \ \Omega$$

The solution impedance is the parallel combination of $Z_{\text{C-soln}}$ (2000 Ω) and R_{soln} (100 Ω), namely 95 Ω. This is small compared to the plate impedance of 6600 Ω, but this latter value is constant and can be subtracted out. A larger AC voltage can be applied, since there is no possibility of any faradaic process occurring, and a large enough drop will appear across the solution to be measured with precision.

Magnetic Induction

The instruments described above utilize capacitive coupling between the sample and the outside world. It is also possible to employ inductive coupling for this purpose. One way to do this is simply to place the solution in its glass container within a coil of wire through which an AC current is flowing. Eddy currents are set up in the solution that will dissipate power in proportion to the conductivity. This was the method used by Jensen and Parrack in the first reported electrodeless conductometric titration [1]. It is also essentially the same that has been adapted to modern electronic circuitry in the most recent high-frequency titrimeter of which the present authors are aware [2].

Another approach is to use two coils, linked together by a loop current induced in the sample. Figure 13-3 shows two instruments designed especially for salinity measurements in sea water [3]. The two coils are wound toroidally around an open plastic tube, the whole lowered by cable into the sea. The current in the primary coil induces a current in the sea water, and this in turn induces a current in the secondary winding.

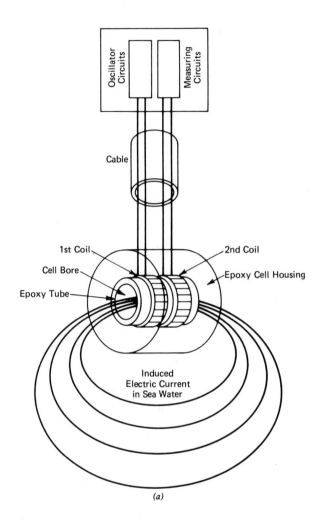

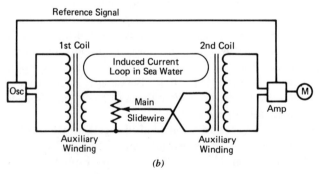

Figure 13-3. (a) A device for the measurement of the total ion concentration (salinity) of sea water. (b) Details of circuit. (Beckman Instruments.)

THEORETICAL CONSIDERATIONS

The quantity actually measured is the conductance G (in seimens). Its dependence on concentration is given approximately by the relation:

$$G = \frac{1}{1000} \cdot \frac{A}{d} \cdot z \Lambda C \qquad (13\text{-}3)$$

where C is the concentration of electrolyte in moles per liter, z is the number of unit charges on each ion, A is the area of each electrode (cm^2), d is the distance separating them (cm), and Λ is a constant known as the *equivalent conductance.* (The factor of 1000 is merely the number of cubic centimeters in one liter.)

If the contribution of each individual ion to the conductance were truly unaffected by the presence of other ions, then Λ would be entirely independent of concentration. This, however, is not the case; Λ decreases slowly with increasing concentration.

The equivalent conductance contains the contributions of all ions of both signs. At high concentrations the interaction between them is reflected in a complicated conductivity function. As the concentration is reduced, it would be expected that the interaction of ions with each other would disappear, and that a linear relation would prevail. This is indeed the case, and at zero concentration (infinite dilution), the following equation holds:

$$G = \frac{1}{1000} \cdot \frac{A}{d} \cdot \sum_i z_i \lambda_i C_i \qquad (13\text{-}4)$$

Here λ_i is the contribution of the ith ion, called the *ionic equivalent conductance.*[†] This relation provides a good approximation at very low concentrations. Table 13-1 gives the equivalent conductances of selected ions extrapolated to zero concentration. The tabular values can be used to calculate the equivalent conductance, $\lambda°$, at infinite dilution. For example, for KCl the value is $73.5 + 76.4 = 149.9$ S $\cdot$ $m^2 \cdot mol^{-1}$. This is a reasonable approximation, since $0.01\ M$ KCl solution shows a measured value of 141.3.

The reciprocal of the geometric factor in the above equations, d/A, is called the *cell constant*, usually evaluated by measurement of a solution of known conductance. It is primarily useful in comparing results obtained with different cells.

INSTRUMENTATION

The Four-Electrode Cell

A direct way to set up a circuit for a DC four-electrode conductance cell is shown [4] in Figure 13-4a. A constant-current power supply forces the desired current

---

[†]The *ionic mobility*, sometimes quoted, is equal to the ionic equivalent conductance divided by the Faraday constant.

TABLE 13-1
Equivalent Ionic Conductance of Ions
At Infinite Dilution
$$(S \cdot cm^2/mol)^a$$

Cations	λ°	Anions	λ°
H^+	349.8	OH^-	198.6
$Co(NH_3)_6^{+++}$	102.3	$Fe(CN)_6^{----}$	110.5
K^+	73.5	$Fe(CN)_6^{---}$	101.0
NH_4^+	73.5	$Co(CN)_6^{---}$	98.9
Pb^{++}	69.5	SO_4^{--}	80.0
La^{+++}	69.6	Br^-	78.1
Fe^{+++}	68.0	I^-	76.8
Ba^{++}	63.6	Cl^-	76.4
Ag^+	61.9	NO_3^-	71.4
Ca^{++}	59.5	CO_3^{--}	69.3
Cu^{++}	53.6	$C_2O_4^{--}$	74.2
Fe^{++}	54.0	ClO_4^-	67.3
Mg^{++}	53.1	HCO_3^-	44.5
Zn^{++}	52.8	$CH_3CO_2^-$	40.9
Na^+	50.1	$HC_2O_4^-$	40.2
Li^+	38.7	$C_6H_5CO_2^-$	32.4
$(n-Bu)_4N^+$	19.5		

^{a}"Mol" refers to moles of ionic charge.

through the outer pair of electrodes, while the voltage appearing between the inner (probe) electrodes is measured with a high-impedance voltmeter. The conductance is then determined as the ratio of current to voltage, $G = I/E$.

The requirement of a constant-current supply can be eliminated if one places the cell in the feedback loop of an operational amplifier [5], as in Figure 13-4b. The reference voltage, E_{ref}, is set at some convenient value, say 1.00 V. Then the amplifier will force just enough current through the cell to produce a drop of exactly 1.00 V between the probe electrodes. The voltage E_{out} across the series resistor R is a measure of the current. The conductance is given by I/E_{ref}, and since E_{ref} is constant and known, the conductance is easily determined.

The Two-Electrode Cell

There are a number of possible connections for a two-electrode conductance cell. The most widely used is the AC *Wheatstone bridge*, Figure 13-5a [6]. Of the four resistors making up the bridge, let R_x be the resistance of the cell with its shunt capacitance C_x, and let R_1 be a series of calibrated resistors in decade steps. In

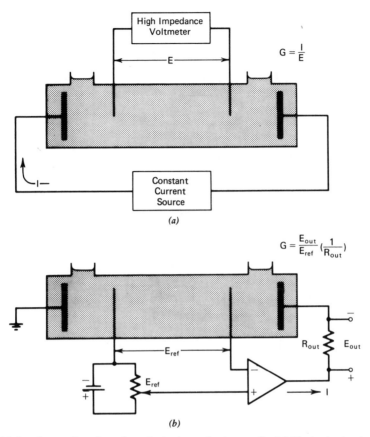

Figure 13-4. Connections for a four-electrode conductance cell. (*a*) The basic circuit. (*b*) As controlled by a single op amp.

operation the bridge is balanced successively for capacitance and resistance by adjustment of C_{var} and R_1, until the meter shows a null reading. At this point:

$$G = \frac{R_1}{R_2 R_3} \quad \text{and} \quad C_{var} = C_x \qquad (13\text{-}5)$$

It is possible to replace the Wheatstone bridge with a somewhat simpler application of an operational amplifier, sometimes called a *pseudo-bridge* [6], Figure 13-5*b*. The same equations hold, and the precision is unchanged. Either the bridge or the pseudo-bridge can be adapted to automatic recording.

Another op amp circuit that makes up in simplicity and convenience for what it lacks in precision is shown in Figure 13-6. The output voltage is:

$$E_{out} = -ERG \qquad (13\text{-}6)$$

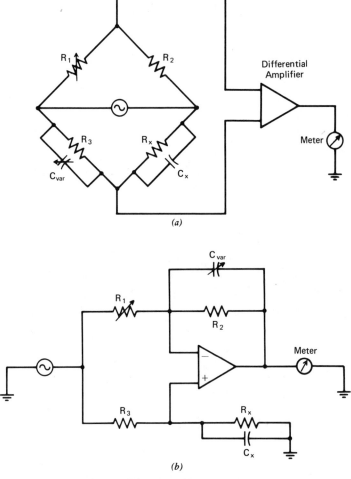

Figure 13-5. (a) A Wheatstone bridge for conductance measurement. (b) A pseudo-bridge. The letter designations correspond in the two diagrams.

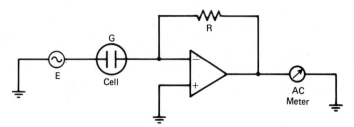

Figure 13-6. Direct-reading conductance meter.

A new approach to conductance measurement was reported in 1973 [7]. In this method, a short pulse of current, I in Figure 13-7, followed by an equal pulse of opposite sign, is passed through a conventional two-electrode cell. The resulting voltage, E, appearing across the cell is also shown in the figure. This voltage is rectified (E_{rect}), and the area beneath this curve is integrated. It is shown mathematically that this integral is strictly proportional to the resistance of the solution.

Electrodeless Cells

The instrument for high-frequency measurements of conductance, called an *oscillometer*, bears little superficial resemblance to the instruments previously discussed. This is because most electronic components have greatly different characteristics at high frequency. Even more significant may be the effect of distributed parameters. Wires themselves show considerable inductance and capacitance with respect to their surroundings. All these factors must be taken into account by the designer and user of apparatus in this field. For further treatment, the reader is referred to [8].

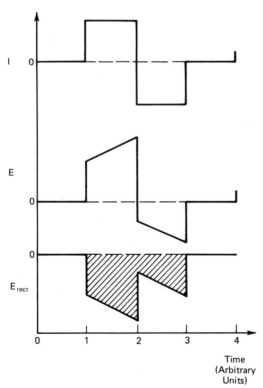

Figure 13-7.   Double-pulse method for conductance measurements. One unit of time may be from 20 μs to 5 min. (*Analytical Chemistry* [7].)

REFERENCES

1. F. W. Jensen and A. L. Parrack, *Ind. Eng. Chem., Anal. Ed.*, **1946**, *18*, 595.

2. A. Scher and C. Yarnitzky, *Anal. Chem.*, **1981**, *53*, 356.

3. Beckman Instruments, Cedar Grove Operations; descriptive literature, **1980**.

4. R. Colton, G. J. Sketchley, and I. M. Ritchie, *J. Chem. Educ.*, **1976**, *53*, 130.

5. J.-K. Chon and W.-K. Paik, *J. Korean Chem. Soc.*, **1976**, *20*, 129.

6. G. W. Ewing, *J. Chem. Educ.*, **1975**, *52*, A239.

7. P. H. Daum and D. F. Nelson, *Anal. Chem.*, **1973**, *45*, 463.

8. J. W. Loveland, in "Treatise on Analytical Chemistry" (I. M. Kolthoff and P. J. Elving, Eds.), Wiley-Interscience, New York, **1963**, Pt. I, Vol. 4, p. 2569ff.

Chapter 14

OPTICAL–ELECTROCHEMICAL METHODS

There are many ways in which optical and electrochemical processes can interact on a common ground. The more important can be classified as follows:

1. The product of an electrochemical reaction (or the electrode itself) is examined by optical means. This is *spectroelectrochemistry*.
2. The product of a photochemical reaction is studied by electrochemical means. This may be called *electrochemical photochemistry*.
3. An electrode produces current as the result of illumination, a process described as the *photoelectrochemical effect*.
4. Light is emitted from an electrochemical cell. The term used to describe this is *electrogenerated chemiluminescence*.
5. *Enhanced Raman spectroscopy* can be classified as a variety of spectroelectrochemistry but deserves separate discussion because of its rather different principle.

To this one might add the use of *in situ* electrochemical preparation of reagents to be used in spectroscopic disciplines, such as ESR and NMR. We shall not treat these cases because the electro- and spectro-operations are essentially independent.

SPECTROELECTROCHEMISTRY

In this type of experiment, use is made of spectroscopy as a probe, to analyze the products of an electrochemical reaction [1]. The process can be implemented by reflecting light from the electrode surface, but more commonly by the use of *optically transparent electrodes*, *OTEs* [2, 3], as illustrated in Figure 14-1*a*. The electrode substrate can be made of glass or quartz for visible or ultraviolet measurements, or of germanium for the infrared region. A very thin platinum, gold,

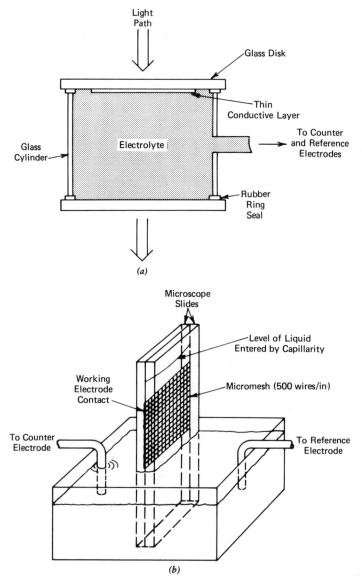

Figure 14-1.   Two types of optically transparent electrodes. (*a*) Glass plate with a conducting surface layer. (*b*) Micromesh electrode.

tin oxide, or other conductive layer is deposited on the surface of the transparent plate.   The requirements for simultaneous transmission of electrical and optical signals are somewhat incompatible, since at small enough thicknesses for light to be transmitted, say 10^{-4} mm, the metal layer exhibits a rather high electrical resistance.

A better alternative is the use of a fine metallic grid or mesh, as shown in Figure

14-1b, that combines reasonable electrical conductivity with good optical transmission (over 50 percent). Such *minigrids*, usually made of gold, with perhaps 50 wires per cm, can be sandwiched between two microscope slides to form an inexpensive thin-layer cell [4, 5]. Another very promising material is reticulated vitreous carbon; a slice of 1 mm thickness exhibits about 20 percent transmittance [6].

The electrical measurement commonly uses voltage-step linear scan voltammetry or coulometry. In the LSV approach, after the application of a voltage step, the current (in the absence of convection) initially follows the Cottrell equation. The optical measurement involves the totality of absorbing material produced and therefore obeys an equation of the type [7]:

$$- \log T = \epsilon \int C_{RED} \, dx \tag{14-1}$$

where T is the transmittance of the product, here considered to be RED, and ϵ is its molar absorptivity. The integral must be taken over the whole cell, even if the product has not yet reached the far wall. For a homogeneous distribution of material, Eq. (14-1) would lead to Beer's law, but in this instance, the concentration is not uniform.

The value of log T can be estimated if we observe that the concentration integral is equal to the current integral divided by nF:

$$\int C_{RED} \, dx = \frac{1}{nF} \int I \, dt \tag{14-2}$$

If we assume that the current is given by the Cottrell equation, we obtain:

$$- \log T = 2\epsilon \left(\frac{Dt}{\pi} \right)^{1/2} C_{OX}^* \tag{14-3}$$

which is quite analogous to Beer's law if we take the quantity $2(Dt/\pi)^{1/2}$ to be an equivalent path length. As long as this distance is smaller than the thickness of the cell, the equation above is likely to be obeyed. When it becomes larger, the process limits itself, as shown in Figure 14-2. Note that the area of the electrode does not enter into the equation, a feature that turns out to be useful in the direct determination of diffusion coefficients.

If the applied potential is scanned, both thin-layer and large body cells produce a peaked current response (Figure 14-3). For the thin-layer case, the peaking is more pronounced. It is assumed that in the plateau region the reaction is complete. If the scan is very slow, equilibrium is approximated at all times, and the Nernst equation applies. Consequently, by determining the point at which the concentrations of OX and RED are equal, it is possible to obtain the formal potential of the couple with considerable confidence. This approach is valid even when the electron-transfer reaction is kinetically inhibited. In this case, the potential could be held constant until equilibrium is achieved, but more conveniently a second redox

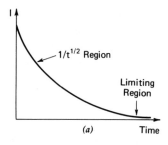

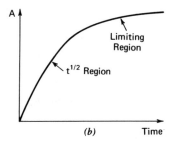

Figure 14-2. Response of a thin-layer cell with an optically transparent electrode. (a) Electrical response; (b) optical response.

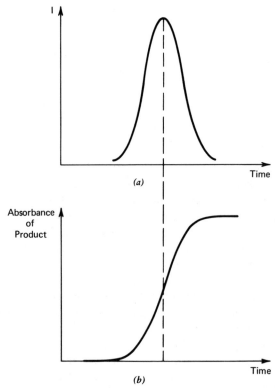

Figure 14-3. Linear scan experiment in a thin-layer cell. (a) Current response; (b) optical absorbance of the product.

Electrode

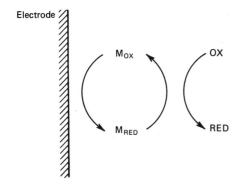

Figure 14-4.   Illustration of a mediator (M) redox process, useful for substances of high molecular weight which oxidize or reduce only slowly.

couple can be added (provided that it does not interfere optically). Thus large, slow-reacting molecules can be made to attain redox equilibrium by means of a fast-reacting redox pair, called a *mediator*, as shown in Figure 14-4. At equilibrium, all four species are involved, and the two redox couples must have the same potential. Consequently, the presence of the mediator does not affect the measurement, aside from the kinetics. The mediator could be ferricyanide for oxidations or methyl viologen† for reductions [8]. Such species as cytochrome-c can be catalyzed in this manner to undergo oxidations or reductions.

Absorption spectra can also be obtained in a reflection mode, as shown in Figure 14-5. Internal reflection measurements can be accomplished using a platinum-covered quartz [9] or, for the infrared, a germanium plate [10]. Any good reflector can be used for specular reflection measurements (Figure 14-5b). Of particular interest is the enhancement obtainable at extremely shallow angles [11]. Thus, for an angle of incidence of 1°, an enhancement by a factor of 100 has been reported. This is important, since the absorbances obtained are as a rule, rather small.

ELECTROCHEMICAL PHOTOCHEMISTRY

The products of a photochemical reaction, for instance, a flash photolysis, can be analyzed conveniently *in situ* by electrochemical means. A certain degree of selectivity can be achieved by potential control, and it is possible to follow the concentration of photochemical products in real time. Careful experiments permit the monitoring of substances that are stable for as little as 0.1 ms or even less, following the flash [12]. The method can be used with conventional voltammetry [13] or with a rotating electrode [14]. A rotating disk system is shown in Figure 14-6. The electrode consists of a platinum ring surrounding a circular quartz window through which the light passes. Photolysis occurs in the same area where there would be product generation in a ring–disk system. The ensemble there-

†Methyl viologen is 1,1′-dimethyl-4,4′-bipyridyl dichloride. See also Reference 23, added in proof.

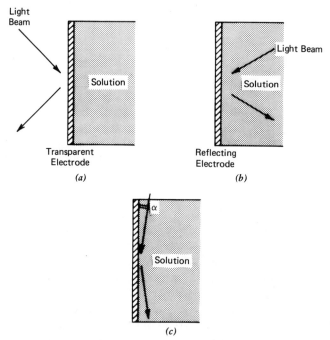

Figure 14-5. (*a*) Internal reflectance in an electrochemical cell. (*b*) Specular reflectance. (*c*) Shallow angle reflection, which increases the path length.

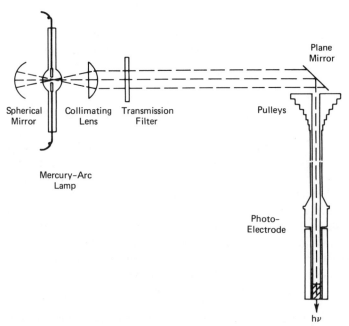

Figure 14-6. Optical system of a photoelectrode. (*Analytical Chemistry* [14].)

fore exhibits similar behavior to ring–disk electrodes. Dropping mercury electrodes have also been used successfully [12, 15].

PHOTOELECTROCHEMISTRY

This describes the situation in which the photochemical reaction occurs at the actual electrode–solution interface, rather than in the bulk of the solution [16, 17]. Such reactions take place primarily at semiconductor electrodes. This must be distinguished from processes where the photolysis occurs in the solution and where the electrode has only an incidental function.

The principle [17] is shown in Figure 14-7a. The electrons in a semiconductor are distributed in valence bands that are filled; at some higher energy there are empty bands called conductance bands. At the interface with the liquid, some distortion occurs, as shown, resulting from interaction with the double layer. The semiconductor depicted is of the n-type in which the principal carriers of electricity (the majority carriers) are electrons. For the current to pass, electrons must

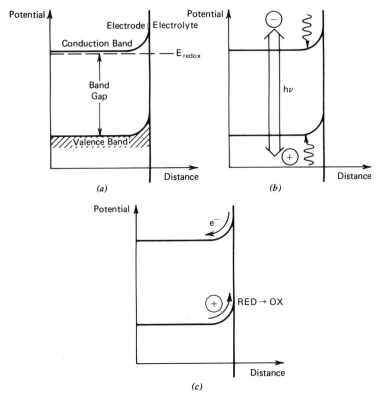

Figure 14-7. (a) Energy levels for an n-type semiconductor–electrolyte interface. For a p-type, the energy bands would curve downwards. (b) The effect of light in separating charges. (c) A redox process. The distance scale is perhaps 100 nm in practical cases.

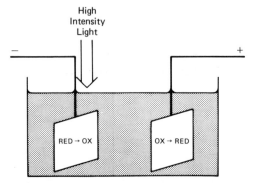

Figure 14-8.   Generation of electricity by illumination.

be promoted across the gap to the conduction band. The required energy can be provided in various ways, but in the present context, the process responsible is photochemical. When a photon of sufficient energy is absorbed at the interface, an electron will be raised to a higher energy level, and the rest of the molecule (left behind as a positive hole) will also go to a higher energy state (plotted downwards in Figure 14-7b). After the excess energy is dissipated thermally, we are left with an electron–hole pair ready to enter spontaneously into a redox process. This consists of the acceptance by the hole of an electron from RED, while the conduction electron moves away to complete the electrical circuit. The oxidation is spontaneous because of the excess energy possessed by the electron–hole pair. Unfortunately, direct recombination of electrons and holes is also possible, so that the efficiency is generally low.

The most direct way to use such a process is for the transformation of light into electricity, as shown in Figure 14-8. The cathode is driven by the reverse process acting on the product of the anodic reaction, so that the cell has no net reaction. Present technology does not permit the construction of efficient cells of this type, in spite of concerted research on the subject.

It is also possible to construct cells in which there is a net reaction and, therefore, a net generation of products. Perhaps the most analytically interesting case is the combination of applied potential and light energy to implement a specific reaction. The light can be regarded as a catalyst, accelerating a process that is already thermodynamically spontaneous. As mentioned in earlier chapters, electrochemistry permits the easy decrease of the ΔG of a reaction by substituting electrochemical for purely chemical free energy. There is, however, no guarantee that the process will actually proceed at a measurable rate, since kinetic barriers are often present. It is therefore very useful to possess an additional kinetic degree of freedom in the design of experiments.

ELECTROCHEMILUMINESCENCE (ECL)

The fundamental process in this category consists in the electrochemical generation of radical anions and cations, X^- and X^+. The recombination of such species leads to the formation of excited states that, in turn, produce light by mechanisms of the type:

$$\text{Singlet route} \begin{cases} X^- + X^+ \longrightarrow {}^1X^* + X \\ {}^1X^* \longrightarrow X + h\nu \end{cases} \tag{14-4}$$

$$\text{Triplet route} \begin{cases} X^- + X^+ \longrightarrow {}^3X^* + X \\ {}^3X^* + {}^3X^* \longrightarrow {}^1X^* + X \\ {}^1X^* \longrightarrow X + h\nu \end{cases} \tag{14-5}$$

The process is carried out in an organic solvent in the absence of water and oxygen. The anion and cation need not be of the same molecular species. They are usually derived from aromatic amines, aromatic hydrocarbons, quinones, and so on. It is even possible to combine the electron donor and acceptor functions in a single molecule.

The generation of the two types of radicals is normally carried out by a double anodic–cathodic step, or better yet, by producing one ion-radical at the disk and the other at a ring in a ring–disk system. Only a few analytical applications are envisioned.

SURFACE-ENHANCED RAMAN SPECTROSCOPY (SERS)

While it is possible and useful to perform spectroelectrochemistry using the conventional Raman effect, it has been shown recently that there is a special mode of operation [21, 22], by which a resonant process enhances the intensity of the Raman spectrum by as much as 10^5 times.

The enhancement is best observed with silver electrodes, but copper and mercury may be used as well. The substance involved must be adsorbed on the surface of the electrode. Pyridine adsorbed on silver has been extensively studied by this method. The process is not well understood, but it appears that the high local fields produced by the roughness of the electrode, as well as image dipoles induced in the metal by adsorbed species, play an important role. The effect, unlike many other spectroscopic phenomena, depends on the potential applied, the nature of the metal, and other *external* parameters. This makes the method an exciting probe capable of producing a wealth of information.

REFERENCES

1. W. R. Heineman, *Anal. Chem.*, **1978**, *50*, 390A.
2. T. Kuwana, R. K. Darlington, and D. W. Leedy, *Anal. Chem.*, **1964**, *36*, 2023.
3. T. Kuwana and W. R. Heineman, *Acc. Chem. Res.*, **1976**, *9*, 241.
4. W. R. Heineman, B. J. Norris, and J. F. Goelz, *Anal. Chem.*, **1975**, *47*, 79.
5. T. P. DeAngelis and W. R. Heineman, *J. Chem. Educ.*, **1976**, *53*, 594.

6.   V. E. Norvell and G. Mamantov, *Anal. Chem.*, **1977**, *49*, 1470.

7.   J. W. Strojek, T, Kuwana, and S. W. Feldberg, *J. Am. Chem. Soc.*, **1968**, *90*, 1353.

8.   F. M. Hawkridge and T. Kuwana, *Anal. Chem.*, **1973**, *45*, 1021.

9.   B. S. Pons, J. S. Mattson, L. O. Winstrom and H. B. Mark, Jr., *Anal. Chem.*, **1967**, *39*, 685.

10. H. B. Mark, Jr. and B. S. Pons, *Anal. Chem.*, **1966**, *38*, 119.

11. R. L. McCreery, R. Pruiksma, and R. Fagan, *Anal. Chem.*, **1979**, *51*, 749.

12. J. R. Birk and S. P. Perone, *Anal. Chem.*, **1968**, *40*, 496.

13. H. Berg, *Collect. Czech. Chem. Commun.*, **1960**, *25*, 3404.

14. D. C. Johnson and E. W. Resnick, *Anal. Chem.*, **1972**, *44*, 637.

15. A. Henglein, *J. Electroanal. Chem.*, **1976**, *9*, 163.

16. P. Delahay and V. S. Srinivasan, *J. Phys. Chem.*, **1966**, *70*, 420.

17. A. J. Bard, *Science*, **1980**, *207*, 139.

18. R. E. Visco and E. A. Chandross, *J. Am. Chem. Soc.*, **1964**, *86*, 5350.

19. D. M. Hercules, *Science*, **1964**, *145*, 808.

20. L. R. Faulkner and A. J. Bard, *J. Electroanal. Chem.*, **1977**, *10*, 1.

21. A. M. Fleishman, P. J. Hendra, and A. J. McQuillen, *Chem. Phys. Lett.*, **1974**, *26*, 163.

22. M. G. Albrecht and J. A. Creighton, *Electrochim. Acta*, **1978**, *23*, 1103.

23. M. L. Fultz and R. A. Durst, *Anal. Chim. Acta.*, **1982**, *140*, 1.

Chapter 15

TECHNIQUES OF MEASUREMENT

We have, in the previous chapters, explored many techniques whereby the concentration of an analyte in solution can be estimated. The purpose of the present chapter is to bring together some procedures of general applicability that can increase the precision of analytical methods and, in some cases, provide additional information of a qualitative nature as well.

As a rule, analysis of a sample involves three steps: (1) preparation of a suitable environment, such as by buffering; (2) the application of a stimulus X and observation of the corresponding response Y; (3) the mathematical processing of Y to give a more convenient quantity, $F(Y)$, usually proportional to the desired concentration of the analyte.

For each electrochemical method, there is a mathematical expression that relates the instrumental response Y to the concentration. Often, this amounts to a direct proportionality over a considerable range, as for example, in polarography, coulometry, and conductometry. In constant-current chronopotentiometry, the relation takes the form of a square-root function, $C = k\sqrt{Y}$. An exponential expression applies for zero-current potentiometry, $C = k_1 \exp(k_2 Y)$.

In most of the methods discussed in this book, the working equation is useful principally for dimensional studies, to show how the variables affect one another. For quantitative analytical objectives, unknowns must be compared to standards. Coulometry is unique in that it allows precise determination of concentration to be made by calculation alone from the measured current–time integral.

COMPARISON WITH STANDARDS

The most direct, but not necessarily the best, procedure is calibration with a standard. In the common case, where Y is proportional to C, three measurements will suffice: a blank B, the standard S, and the sample X, all in the same chemical environment. Calculation then follows the formula:

$$c_X = c_S\left(\frac{Y_X - Y_B}{Y_S - Y_B}\right) \qquad (15\text{-}1)$$

This method is most accurate if the concentrations of S and X are nearly equal. If this turns out not to be true, a second standard should then be prepared, duplicating more closely the estimated value of c_X. A more complete procedure is to make instrumental measurements on a series of standard solutions of the analyte to cover the expected range.

The maximum span over which reliable measurements can be made is known as the *dynamic range*. Its lower limit is established by random fluctuations (noise) in the instrumental response. The ratio of signal to noise (S/N) should be at least 2 for a measurement to be just observable, but a larger ratio is required for high precision. The upper limit of the range is imposed by some kind of saturation phenomenon, for example, the extent of solubility of the analyte or a limit in the ability of the detector to respond.

A plot of analytical data, such as Figure 15-1, reveals much of interest about the method: (1) the mathematical function (though this is usually known in advance). The question to be asked is: Are the data approximately linear with respect to Y, to $\log Y$, to Y^2 or some other function? This may mean that the data must be replotted against $F(Y)$ to give a linear plot. (2) The presence of a background. This is indicated by a nonzero intercept on the ordinate axis, which would be found, for example, in DC polarography if correction for residual current were omitted. (3) Curvature as the plot is extended to higher concentrations. This suggests approach to the limit of the dynamic range of the technique. (4) Scatter of the data points, a rough indication of precision.

A danger in the use of standards lies in possible *matrix effects*, a change in sensitivity due to the presence of extraneous material in the sample. This is a likely occurrence if the standards contain only A, while the unknown contains variable amounts of other constituents.

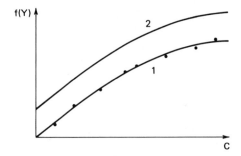

Figure 15-1. Typical calibration curves. Curve 2 shows the case for a nonzero intercept on the ordinate axis. Hypothetical experimental points are shown for Curve 1.

STANDARD ADDITION OR SUBTRACTION

Undesirable matrix effects can be diminished or perhaps eliminated entirely by the application of the method of standard addition, in which a portion of standard is added directly to the sample. Consider first the case where the instrument response, Y, is directly proportional to concentration after correction for any background. We can write, using subscript zero for the original solution,

$$Y_0 = k(M_0/V_0) \tag{15-2}$$

where concentration is represented as the number of moles M in volume V. Suppose now that we add a portion of the standard solution containing M_1 moles in volume V_1. Rewriting Eq. (15-2) for the resulting solution gives:

$$Y_1 = k\left(\frac{M_0 + M_1}{V_0 + V_1}\right) \tag{15-3}$$

This can be simplified if V_1 is kept small with respect to V_0, as by using a sufficiently concentrated standard, so that:

$$Y_1 = \frac{k}{V_0}(M_0 + M_1) \tag{15-4}$$

Eliminating (k/V_0) between Eqs. (15-2) and (15-4), and solving for M_0, gives:

$$M_0 = M_1\left(\frac{Y_0}{Y_1 - Y_0}\right) \tag{15-5}$$

This procedure eliminates all instrumental variables, represented by the quantity k. A determination of M_1 and, hence, of C can then be made, given knowledge of the volume.

It is essential that the measurements with and without the standard be made in the presence of identical amounts of all other components of the solution. It is not prudent to assume that the matrix effects are completely eliminated, however, so the procedure should be repeated with additional increments of standard to give readings Y_2 and Y_3. The results can be plotted as in Figure 15-2, and M_0 evaluated graphically.

For the exponential case, where $Y = Y' + K \ln (M/V)$, a similar treatment, incrementing M, gives:

$$M_0 = M_1\left[\exp\left(\frac{Y_1 - Y_0}{k}\right) - 1\right]^{-1} \tag{15-6}$$

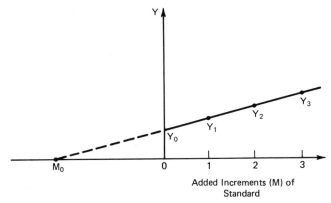

Figure 15-2.   Graphical determination of M_0 from standard addition data. This plot can serve to check the linearity of response.

When applied to potentiometry, k becomes the familiar factor RT/nF. A graphical solution in this case must be made on semilogarithmic paper (*Gran's plot paper* [1]), on which systems exhibiting exponential response give graphs similar to that in Figure 15-2.

Comparable expressions to Eqs. (15-5) and (15-6) can be derived for other functions of concentrations.

It can be shown [2] that best precision can be obtained when the added increment is somewhat larger than the amount of analyte initially present. A more complex treatment must be drawn upon in the presence of strong interferences with the desired analyte [3].

Sometimes a procedure called *standard subtraction* is applicable, in which one adds a known amount of a reagent that will effectively remove a corresponding amount of analyte from the reaction by precipitation or complexation. The mathematical relations are the same as for standard addition, with appropriate changes of sign.

It is important to reiterate that, in any comparison with standards, all conditions must be held uniform. This is sometimes difficult or impossible to do, but is nevertheless required for highest precision. The ionic strength and the concentration of complexing agents, for example, should not be allowed to vary when a standard addition is made. Also, it must be remembered that the accuracy of a result can be no better than the accuracy with which the standard solution is prepared.

TITRIMETRY

Titration can be considered to be a modification of the familiar method of decreasing relative error by taking replicate determinations. The precision of a single measurement can be improved by a factor of 4 by repeating it for a total of $4^2 = 16$

replicates. If the analyte be titrated with 16 volumetric additions, these 16 observations are spread out over a series of concentrations, and drawing the best line through them will increase the precision by a corresponding factor.

A titration may, of course, provide a measure of a different quantity than that obtained from a single point, whether replicated or not. A prime example is a weak acid; a simple pH measurement gives the H^+ ion activity only, whereas a titration, monitored by the same pH meter, tells us the total quantity of acid initially present, as well as the H^+ activity at any point during the titration.

Titrimetry can be defined as the science and art of determining the amount of a substance by reaction with an equivalent quantity of a reagent added in measured amounts. Two types of observations must be made in any titration: (1) the equivalence point must be identified, and (2) the quantity of reagent needed to reach that point must be measured.

In electroanalytical applications, the measurement of quantity added in a titration is ordinarily either volumetric or coulometric. Endpoint detection can be achieved through monitoring the course of the reaction by a variety of techniques described in this text.

Endpoint Detection

Zero-Current Potentiometry. The course of a titration reaction can be followed by observation of the potential of an indicator electrode that senses either the analyte or the reagent. Because of the logarithmic form of the Nernst equation, the titration curve has the familiar wave shape, the inflection point corresponding to the endpoint. Details of such curves can be found in texts on quantitative chemistry and need not be repeated here.

Satisfactory titrations can often be obtained with electrode pairs that thermodynamic theory would predict to be useless. An example is the combination of tungsten and platinum electrodes. Both are "inert," and should sense the same redox potential. However, equilibrium is attained much more slowly at tungsten than at platinum, so that one lags in response behind the other. This system shows a sudden break at the endpoint if the reagent is added at a constant rate, as in an automatic or coulometric titrator. It can be used for both redox and acid–base titrations.

The instrumentation for potentiometric titrimetry can be rather simple, since exact potentials are not required, and only the abrupt change at the endpoint is desired. Thus, calibration of a pH meter with standard buffer can be eliminated without error.

To exemplify this and other methods, it is convenient to select a particular redox reaction in which all species involved are soluble, and both electrode reactions act reversibly. For this purpose we make the same choice we did in Chapter 10, the oxidation of ferrous iron by quadrivalent cerium:

$$Fe(II) + Ce(IV) \longrightarrow Fe(III) + Ce(III) \tag{15-7}$$

In 1 M H_2SO_4, the formal potentials are (vs SCE):

$$Fe(III) + e^- \longrightarrow Fe(II) \qquad E^{\circ\prime} = +0.44 \text{ V} \qquad (15\text{-}8)$$

$$Ce(IV) + e^- \longrightarrow Ce(III) \qquad E^{\circ\prime} = +1.19 \qquad (15\text{-}9)$$

In the next few figures, we show idealized voltammograms corresponding to four stages in this reaction. In Figure 15-3 is shown the interpretation of this system in

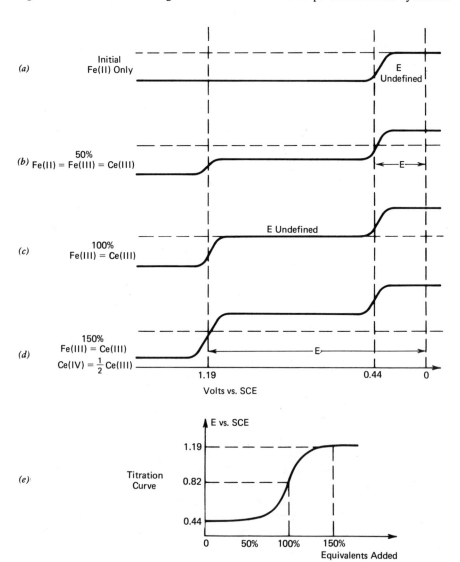

Figure 15-3. Potentiometric titration. (a through d): Voltammograms at various stages of the reaction, as marked; the ordinate is current. (e) The resulting titration curve.

terms of potentiometry. Superimposed on each curve is an indication of the equilibrium potential that would be observed, relative to SCE, at zero current. This potential is determined by the intersection of each curve with the zero-current line. The potential cannot be established unequivocally in part (a), prior to starting the titration, since it is assumed that no ferric ion is present. At the equivalence point (c), the potential would theoretically be at the midpoint between 0.44 and 1.19, namely 0.82 V, but practically could be anywhere within a rather wide range, since the exact location of the curve depends on ubiquitous minute currents due to side reactions. In (b), where one-half equivalent of reagent has been added, and at (d), where another half equivalent beyond the endpoint is present, the potential is firmly fixed. The corresponding potentiometric titration curve is shown at (e).

Bipotentiometric Titration. This involves two platinum electrodes in a stirred solution, with a small current, say 5 mA, passed through them. The resulting potential difference is to be followed during the course of the reaction. In Figure 15-4, the same voltammograms as before are shown and, in addition, two horizontal lines corresponding to the same small current passing through both electrodes. A potential will appear between the two electrodes corresponding to the distance between the intersections of the two current lines with the voltammogram. It will be seen that this potential, ΔE, will be very small at condition (b), where both Fe(II) and Fe(III) are present, and at (d) in the presence of both oxidation states of cerium. At the equivalence point (c), ΔE will be quite large, about 0.75 V. Prior to the start of the titration, at (a), ΔE will be large, terminating at some point, not shown, corresponding to the reduction of another component of the solution, probably H^+ ion. The titration curve is shown at (e).

Biamperometric Titration. Similarly, we might impress a voltage ΔE of, say, 50 mV, across the same two electrodes and monitor the current, as depicted in Figure 15-5. A current can only flow between these electrodes when the interval ΔE can lock onto a point where the voltammetric curve crosses the zero-current level so that one boundary of the interval intersects the curve above, the other at an equal distance below, the zero line. Accordingly, no current can flow at conditions (a) and (c), but current will flow at (b) and (d). Again, (e) shows the resulting titration curve. This method, formerly known as a "dead-stop" titration is widely used because of its simplicity.

Amperometric Titration with a Potentiostat. In this case, the current is controlled by one electrode only, the working electrode, since the auxiliary electrode is without effect. An example is illustrated in Figure 15-6, from which one can visualize the importance of choosing the proper potential; two of the possible titration curves are useless. Amperometric titrations can be carried out with a variety of other electrodes, including the DME (with the latter, formerly referred to as polarometric titration). DME titrations are inconvenient, because of the restriction

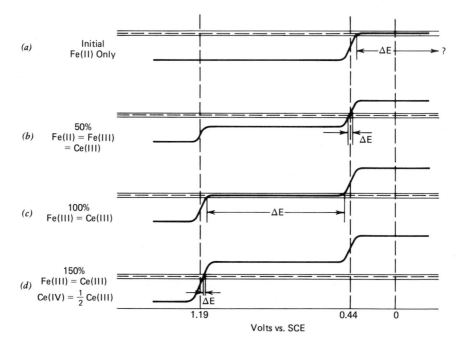

(a) Initial Fe(II) Only

(b) 50% Fe(II) = Fe(III) = Ce(III)

(c) 100% Fe(III) = Ce(III)

(d) 150% Fe(III) = Ce(III), Ce(IV) = ½ Ce(III)

1.19   0.44 0

Volts vs. SCE

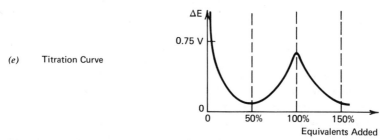

(e) Titration Curve

Figure 15-4. Differential potentiometric titration with two indicator electrodes. Curves as in Figure 15-3.

against stirring, but have the advantages of good reproducibility and sensitivity characteristics of this electrode.

Conductometric Titration. Since conductance relates to the total ionic content of a solution, it can be used to follow any reaction that is accompanied by a change in this quantity. This is true of many titration reactions, including acid-base, precipitation, and complexation, but it is not true of some others. The cerium–iron reaction [Eq. (15-7)], for example, cannot be followed conductometrically, because one ion is simply replaced by another, without the formation of an insoluble

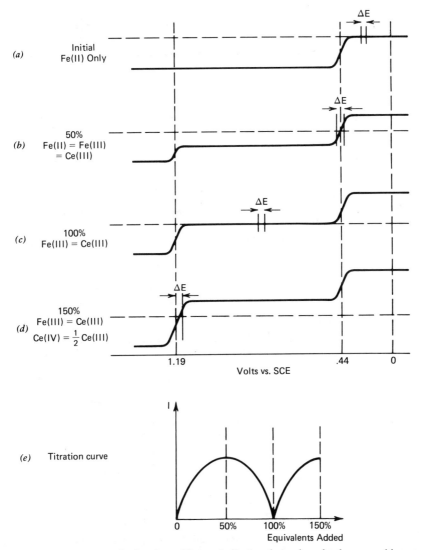

Figure 15-5.   Amperometric titration with two indicator electrodes, also known as biampero-metric titration. Curves as in Figure 15-3.

or un-ionized species.   Some redox reactions [Fe(II) titrated by MnO_4^- in the presence of H_2SO_4 is an example] do show a conductometric endpoint, but with relatively poor precision because of the large amount of unreacting electrolyte present, that tends to swamp out the effect one wishes to observe.

High-frequency conductometry (*oscillometry*) is useful for titration, particularly with solutions that would be incompatible for one reason or another with platinum electrodes.

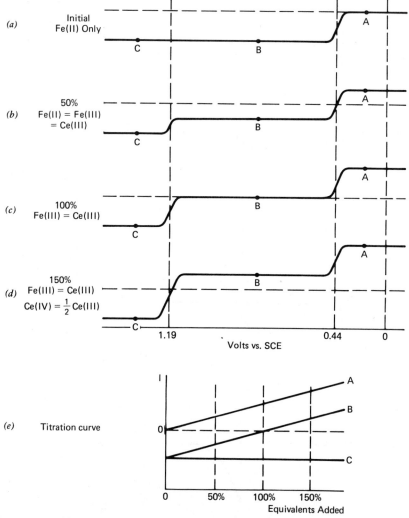

Figure 15-6.   Three-electrode amperometric titration with a platinum working electrode, with stirring. Curves as in Figure 15-3.

Miscellaneous. A number of other electroanalytical techniques have been proposed for endpoint detection, but have not been found to offer significant advantages. Examples are chronopotentiometric [4] and coulostatic [5] titrations.

Constant-Potential Titrimetry

Another approach to potentiometric titration is the measurement of the amount of titrant needed to maintain an indicator electrode at a constant potential. The

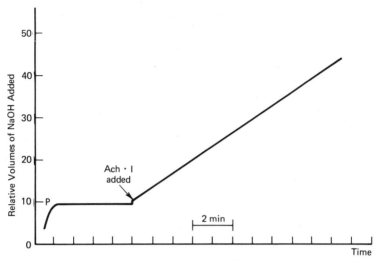

Figure 15-7.   Titration of cholinesterase at constant pH.   (*Acta Pharmacologica et Toxicologica* [6], redrawn.)

titration curve then becomes a plot of volume of standard solution added as a function of time.

This procedure has been used rather extensively in the field of enzymology. For example, the enzyme cholinesterase acts to decompose acetylcholine, producing acetic acid in the process [6]. The enzyme is highly sensitive to pH, and the medium must be held very close to pH 7.4 for the reaction to proceed optimally.

This method makes use of pH meter connected through a feedback loop to control a pump supplying NaOH solution as needed to maintain the pH. A pH meter and associated equipment used in this manner is called a *pH-stat*. Figure 15-7 shows an example of a curve by which the cholinesterase activity of a sample of animal tissue was determined [6]. The pH was adjusted to 7.4 (at point *P* on the curve), and held there for about 10 minutes to establish a base line. Then, an excess of acetylcholine iodide (Ach-I) was added, whereupon the inflow of NaOH commenced.

This technique is useful in following any reaction that evolves or consumes either acid or base, whether for titration or for kinetic studies [7].

Coulometric Titration

As discussed in Chapter 10, reactive species can be generated electrolytically to react with an analyte, up to an endpoint identifiable by an appropriate means. This procedure definitely falls within the concept of titration.

A large variety of reagents can be generated electrically so that all types of titration—acid-base, redox, precipitation, complexation—can be implemented coulometrically. A partial list of reagents and their precursors is given in Table

15-1 [8]. References [8] and [9] include many coulometric titrations that have been reported in the literature.

Coulometric titrimetry offers a number of advantages over volumetric methods, largely in terms of convenience. Electrical measurements, capable of high precision, obviate the need for primary standards for routine use. The method is best suited to small samples (say up to 50 μg of analyte) where the relative standard deviation typically is less than 0.5 percent [10]. Large samples involve large values of $Q = I\Delta t$, which, in turn, require either a sizable current, and thus the likelihood of side reactions, or a long time. As a rule of thumb, about 0.2 milliequivalent is a reasonable upper limit; this corresponds to about 30 mA flowing for a period of 10 minutes.

TABLE 15-1
Coulometric Reagents
and their Precursors [8][a]

Reagent	Precursor
Ag^+	Ag (silver anode)
Ag(II)	$AgNO_3$
Br_2	KBr (pH $<$ 5)
BrO^-	KBr (pH 8–8.5)
Ce(IV)	$Ce_2(SO_4)_3$
Cl_2	KCl (pH $<$ 1)
Cu(I)	$CuSO_4$
Fe(II)	$Fe_2(SO_4)_3$
$Fe(CN)_6^{---}$	$Fe(CN)_6^{----}$
$Fe(CN)_6^{----}$	$Fe(CN)_6^{---}$
$Fe(EDTA)^{--}$	$Fe(EDTA)^-$
H^+	$H_2O + Na_2SO_4$
Hg(I)	Hg (pH 2–3)
Hg(II)	Hg (pH 9–12)
$HSCH_2CO_2^-$	$Hg(SCH_2CO_2)_2$
I_2 (I_3^-)	KI (pH $<$ 9)
Mn(III)	$MnSO_4$
OH^-	$H_2O + Na_2SO_4$
$S_2O_4^{--}$	$NaHSO_3$ (pH 3–5)
Sn(II)	$SnCl_4$
Ti(III)	$TiOSO_4$
U(V)	$UO_2(ClO_4)_2$
U(IV)	UO_2SO_4
V(IV)	$NaVO_3$

[a]For additional requirements for each entry, see Reference 8.

REFERENCES

1. Anon., *Specific Ion Electrode Newsletter* (Orion Researches, Inc.), **1970**, *2*, 49.

2. K. L. Ratzlaff, *Anal. Chem.*, **1979**, *51*, 232.

3. B. E. H. Saxberg and B. R. Kowalski, *Anal. Chem.*, **1979**, *51*, 1031.

4. C. N. Reilley and W. G. Scribner, *Anal. Chem.*, **1955**, *27*, 1210.

5. R. W. Sorenson and R. F. Sympson, *Anal. Chem.*, **1967**, *39*, 1238.

6. J. Jensen-Holm, H. H. Lausen, K. Milthers, and K. O. Møller, *Acta Pharmacol. Toxicol.*, **1959**, *15*, 384.

7. D. M. Rackham, J. B. Chakrabarti, and G. L. O. Davies, *Talanta*, **1981**, *28*, 329.

8. P. S. Farrington, in "Handbook of Analytical Chemistry" (L. Meites, Ed.), McGraw-Hill, New York, **1963**, p. 5-187 ff.

9. J. T. Stock, *Anal. Chem.*, **1980**, *52*, 1R, and previous reviews in this series.

10. G. W. Ewing, *Am. Lab.*, **1981**, *13(6)*, 16.

Chapter 16

SOME ASPECTS OF DIFFUSION PHENOMENA

Diffusion theory has been discussed in bits and pieces in various chapters in this book, but there is need for a unified treatment and an in-depth presentation of certain features. This chapter is designed to be self-sufficient, hence some duplication of material presented earlier will be encountered. A generalized equation for dynamic methods will be derived, followed by a short presentation of the fractional calculus as applied to electrochemistry.

DIFFUSION TRANSPORT

In any liquid solution, all particles, both of solvent and of solutes, are in constant chaotic motion. In a homogeneous, unstirred liquid at equilibrium, this motion is completely random, and no net transport of any species from one region to another will occur. Consider, however, the effect of perturbing this dynamic equilibrium by electrodeposition of a metal as in Figure 16-1a. As a consequence, in the region labelled "S," near the electrode surface, the average concentration of reducible cations is less than in the bulk of solution, "B." Hence, the probability of such ions passing in the direction from B to S is manifestly greater than from S to B. The net number, N, of ions (in moles) passing through each unit area, A, of the plane of X_1, per second, is called the *flux*, Φ:

$$\Phi = \frac{1}{A} \cdot \frac{dN}{dt} \tag{16-1}$$

We can see immediately that the flux will increase with the difference in concentration between B and S; it can be assumed to be directly proportional to the slope of the concentration profile:

$$\Phi = -D\frac{dC_{OX}}{dx} \tag{16-2}$$

where D is a constant of proportionality, the *diffusion coefficient*, carrying the units $(cm^2\ s^{-1})$. (The negative sign appears because the net diffusion is toward the *smaller* concentration.)

However, we are really interested in flux and concentration, both as functions of two variables, the distance x from the electrode surface and the time t, and so we can recast Eq. (16-2) as:

$$\Phi(x,t) = -D\frac{\partial C_{OX}(x,t)}{\partial x} \tag{16-3}$$

This relation, known as *Fick's first law of diffusion*, was originally stated in 1855.

Since two variables are involved, it is useful to derive a second diffusion equation. This can be done by considering the accumulation of matter at a given point.

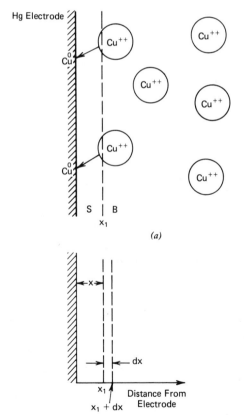

(a)

(b)

Figure 16-1. (*a*) Cu(II) ions being discharged at a mercury electrode. The reduced copper atoms dissolve in the mercury and diffuse away from the interface. (*b*) Differential definitions.

(The variation of concentration in the Y and Z directions is considered to be nil.) The change in concentration of a thin slab of thickness dx parallel to the electrode, with respect to time, will depend on the difference between the incoming and outgoing fluxes, $d\Phi$. Thus, we can write:

$$\frac{\partial \Phi}{\partial x} = \frac{\partial C_{OX}(x,t)}{\partial t} \tag{16-4}$$

and

$$\frac{\partial C_{OX}(x,t)}{\partial t} = \frac{\partial}{\partial x}\left(-D\frac{\partial C_{OX}(x,t)}{\partial x}\right) \tag{16-5}$$

from which

$$\frac{\partial C_{OX}(x,t)}{\partial t} = -D\frac{\partial^2 C_{OX}(x,t)}{\partial x^2} \tag{16-6}$$

This is *Fick's second law of diffusion*.

In order to solve the differential equations, we must assign initial and boundary conditions shown by the following considerations. Assume that before we start an experiment the concentration everywhere is uniform, which means that for time $t = 0$, $C_{OX}^* = C_{OX}^s$. Also, the effect of the electrolysis at any time can extend only some finite distance from the electrode, leaving the rest of the solution unchanged. Thus, we can write a combined equation for the two boundary conditions:

$$C_{OX}(x,0) = C_{OX}^* = C_{OX}(\infty, t) \tag{16-7}$$

We wish to solve Eq. (16-6) with its boundary conditions in such a way as to give us a generalized equation from which the working relations for specific electrochemical regimes can be derived. For convenience we will consider a reduction such that both OX and RED are soluble, either in aqueous solution or in mercury.

The flux of ions resulting from diffusion [Eq. (16-3)] must exactly balance the flux of electrons passing through the electrode [Eq. (16-1)]. Hence, we can write:

$$\Phi = \frac{1}{A}\cdot\frac{dN}{dt} = \frac{I(t)}{nFA(t)} = -D\left(\frac{\partial C_{OX}}{\partial x}\right)^s \tag{16-8}$$

thus combining the laws of Fick and Faraday. To ensure generality, we are considering both current and electrode area to be time-dependent. However, for simplicity, we shall omit notation of the variables (x,t) unless required for emphasis.

Since the concentration is a function of time and distance from an electrode, it follows that:

$$dC_{OX} = \left(\frac{\partial C_{OX}}{\partial x}\right)_t dx + \left(\frac{\partial C_{OX}}{\partial t}\right)_x dt \qquad (16\text{-}9)$$

In order to treat mathematically the effect of changing electrode area, it is necessary to specify the form of this change. We will assume that its variation can be described by a power-of-time function

$$A(t) = at^P \qquad (16\text{-}10)$$

where a and p are positive constants. Note than an electrode of fixed area constitutes a special case where $p = 0$. The DME with constant inflow of mercury corresponds to $p = \frac{2}{3}$, and it can readily be shown that in this case, $a = (36\pi)^{1/3} \rho^{-2/3} m^{2/3}$, where ρ is the density of mercury and m is its rate of flow.

The effect of an increase in area is to create a convective mechanism that must be added to the diffusive term:

$$\frac{dC_{OX}}{dt} = \left(\begin{matrix}\text{Diffusive}\\\text{transport}\end{matrix}\right) + \left(\begin{matrix}\text{Convective}\\\text{transport}\end{matrix}\right) \qquad (16\text{-}11)$$

We can estimate the effect of this convection by making use of the principle of conservation of matter. We define a surface parallel to the expanding electrode, at a distance x_1 from it, so that the volume $x_1 A$ contained between them is constant, i.e.:

$$d(x_1 A) = 0 \qquad (16\text{-}12)$$

or

$$A dx_1 + x_1 dA = 0 \qquad (16\text{-}13)$$

Substituting for A its equivalent from Eq. (16-10) gives:

$$at^P dx_1 = -ax_1 pt^{(p-1)} dt \qquad (16\text{-}14)$$

and

$$\frac{dx_1}{dt} = -\frac{px_1}{t} \qquad (16\text{-}15)$$

The implication of this equation is that a molecule sitting on the plane at x_1 is actually travelling toward the electrode with a speed dx_1/dt. This motion is the convective transport mentioned in Eq. (16-11). Note that this arrangement is valid for *any* plane, so that the subscript ($_1$) can be dropped.

If the effect of diffusion be neglected for the moment, we can state that the concentration of OX at the moving surface defined above is constant: $dC_{OX} = 0$, and Eq. (16-9) can be rewritten as:

$$\left(\frac{\partial C_{OX}}{\partial t}\right)_x = -\left(\frac{\partial C_{OX}}{\partial x}\right)_t \cdot \frac{dx}{dt} \tag{16-6}$$

The value of dx/dt from Eq. (16-15) can then be inserted to give:

$$\left(\frac{\partial C_{OX}}{\partial t}\right)_x = \frac{px}{t}\left(\frac{\partial C_{OX}}{\partial x}\right)_t \tag{16-17}$$

To this expression, which gives the effect of electrode expansion only, must be added the effect of diffusion as given by Eq. (16-6):

$$\left(\frac{\partial C_{OX}}{\partial t}\right)_x = \frac{px}{t} \cdot \left(\frac{\partial C_{OX}}{\partial x}\right)_t + D \cdot \left(\frac{\partial^2 C_{OX}}{\partial x^2}\right)_t \tag{16-18}$$

The two conditions previously mentioned, must be laid on this expression, namely:

$$C_{OX}(x,0) = C_{OX}^* = C_{OX}(\infty,t) \tag{16-19}$$

So far we have concerned ourselves only with the oxidized species, OX, but for a complete picture we must include the product of reduction, RED. Clearly, since we start with only OX present, we can say that:

$$C_{RED}(x,0) = 0 \tag{16-20}$$

If we assume RED to be soluble, then Eqs. (16-9), (16-18), and (16-19) will hold for it as well as for OX, but in general, the diffusion coefficients will differ. This can be taken into account through the relation [1]:

$$C_{OX}^s(t) = C_{OX}^* - \left(\frac{D_{RED}}{D_{OX}}\right)^{1/2} C_{RED}^s(t) \tag{16-21}$$

The General Equation

An additional constraint to be taken into account is the stoichiometric condition inherent in the equation OX + ne^- ⟶ RED, namely:

$$\Phi_{OX} = -\Phi_{RED} = \frac{I}{nFA} \tag{16-22}$$

where I/nFa is the flux of electrons in terms of mol $\cdot$ cm^{-2} s^{-1}, which is also equal to the specific reaction rate, v/A.

As seen in Chapter 2, v/A is given by the equation:

$$\frac{v}{A} = k^{\circ} \left(C_{OX}^{s} \exp\left\{-\frac{\alpha nF}{RT}(E - E^{\circ\prime})\right\} \right.$$
$$\left. - C_{RED}^{s} \exp\left\{\frac{(1 - \alpha)nF}{RT}(E - E^{\circ\prime})\right\} \right) \tag{16-23}$$

where $E^{\circ\prime}$ is the formal potential of the redox couple, and all other symbols have their usual significance.

This expression can be converted to a more convenient form by changing the reference potential from $E^{\circ\prime}$ to $E_{1/2}$. This can be accomplished through the following property of exponentials:

$$\exp(a - c) = \exp(b - c) \cdot \exp(a - b) \tag{16-24}$$

which permits the transformation of Eq. (16-23) into:

$$\frac{v}{A} = k^{\circ} \left(C_{OX}^{s} \exp\left\{-\frac{\alpha nF}{RT}(E_{1/2} - E^{\circ\prime})\right\} \right.$$
$$\cdot \exp\left\{-\frac{\alpha nF}{RT}(E - E_{1/2})\right\}$$
$$- C_{RED}^{s} \exp\left\{\frac{(1 - \alpha)nF}{RT}(E_{1/2} - E^{\circ\prime})\right\}$$
$$\left. \cdot \exp\left\{\frac{(1 - \alpha)nF}{RT}(E - E_{1/2})\right\} \right) \tag{16-25}$$

Recall that $E_{1/2}$ is defined as:

$$E_{1/2} = E^{\circ\prime} + \frac{RT}{nF} \ln\left(\frac{D_{RED}}{D_{OX}}\right)^{1/2} \tag{16-26}$$

from which, by algebraic manipulation:

$$\left(\frac{D_{RED}}{D_{OX}}\right)^{\alpha/2} = \exp\left\{\frac{\alpha nF}{RT}(E_{1/2} - E^{\circ\prime})\right\} \tag{16-27}$$

Consequently, Eq. (16-25) can be rewritten as:

$$\frac{v}{A} = k^{\circ} \left[C_{OX}^{s} \left(\frac{D_{RED}}{D_{OX}}\right)^{-\alpha/2} \exp\left\{-\frac{\alpha nF}{RT}(E - E_{1/2})\right\} \right.$$
$$\left. - C_{RED}^{s} \left(\frac{D_{RED}}{D_{OX}}\right)^{(1-\alpha)/2} \exp\left\{\frac{(1 - \alpha)nF}{RT}(E - E_{1/2})\right\} \right] \tag{16-28}$$

or

$$\frac{v}{A} = k° \left(\frac{D_{RED}}{D_{OX}}\right)^{-\alpha/2} \left(C_{OX}^s \exp\left\{-\frac{\alpha nF}{RT}(E - E_{1/2})\right\}\right.$$
$$\left. - C_{RED}^s \left(\frac{D_{RED}}{D_{OX}}\right)^{1/2} \exp\left\{\frac{(1-\alpha)nF}{RT}(E - E_{1/2})\right\}\right) \qquad (16\text{-}29)$$

This equation can be simplified by the introduction of a new rate constant:

$$k_{1/2} = k° \left(\frac{D_{RED}}{D_{OX}}\right)^{-\alpha/2} \qquad (16\text{-}30)$$

and by replacing $(D_{red}/D_{ox})^{1/2}$ by its equal from Eq. (16-21) to give:

$$\frac{v}{A} = k_{1/2} \left(C_{OX}^s \exp\left\{-\frac{\alpha nF}{RT}(E - E_{1/2})\right\}\right.$$
$$\left. - (C_{OX}^* - C_{OX}^s) \exp\left\{\frac{(1-\alpha)nF}{RT}(E - E_{1/2})\right\}\right) \qquad (16\text{-}31)$$

Equation (16-31) can be solved for C_{OX}^s by straightforward, albeit lengthly, algebra, to give:

$$C^s = \frac{C^* + \dfrac{v}{Ak_{1/2}} \exp\left\{-\dfrac{(1-\alpha)nF}{RT}(E - E_{1/2})\right\}}{1 + \exp\left\{-\dfrac{nF}{RT}(E - E_{1/2})\right\}} \qquad (16\text{-}32)$$

This is the general equation we have been seeking. It is applicable to both reversible and irreversible electrode reactions, provided only that they are under diffusion control. For reversible process, the value of $k_{1/2}$ becomes very large, and the term involving it drops out, and we have:

$$C_{OX\,(rev)}^s = \frac{C_{OX}^*}{1 + \exp\left\{-\dfrac{nF}{RT}(E - E_{1/2})\right\}} \qquad (16\text{-}33)$$

On the other hand, for a process that is completely irreversible, $k_{1/2}$ is very small, and no significant reaction takes place until $(E - E_{1/2})$ is large, so that 1 in the denominator can be neglected. Also, the term $C_{OX}^* [1 + \exp\{-(nF/RT)(E - E_{1/2})\}]^{-1}$ becomes small, and the equation reduces to:

$$C_{OX\,(irrev)}^s = \frac{v}{AK_{1/2}} \exp\left\{\frac{\alpha nF}{RT}(E - E_{1/2})\right\} \qquad (16\text{-}34)$$

It is possible to combine Eqs. (16-32) and (16-18) to produce a relation between potential, current, and bulk concentration. This is a complicated expression, best presented in Laplace notation, as shown, e.g., by Oldham [1], who indicates that substitution of suitable conditions can lead to derivations of many of the equations presented in earlier chapters of this book, thus emphasizing their close relationships. Specifically, Oldham derives, among others:

1. The Cottrell equation, Eq. (4-6)
2. The Heyrovský–Ilkovič equation, Eq. (4-17)
3. The Karaoglanoff equation, Eq. (9-8)
4. The Sand equation, Eq. (9-10)

together with their equivalents for irreversible processes.

SEMI-INTEGRAL AND SEMIDIFFERENTIAL TECHNIQUES

Oldham and co-workers [1, 2] in 1969–1970 first pointed out the possible significance in electrochemistry of the semi-integral and semidifferential mathematical operators. For detailed background, see [3]. These operators are represented by the notation: $(d^{\pm 1/2}/dt^{\pm 1/2})f(t)$, wherein the positive exponent applies to semi-differentiation and the negative to the inverse, semi-integration. The two operators, as applied to currents, generate new functions of considerable interest:

$$\mu(t) \equiv \frac{d^{-1/2}}{dt^{-1/2}} I(t) \tag{16-35}$$

$$\epsilon(t) \equiv \frac{d^{1/2}}{dt^{1/2}} I(t) \tag{16-36}$$

The units for $\mu(t)$ are $A \cdot s^{1/2}$ and for $\epsilon(t)$, $A \cdot s^{-1/2}$. (The name "amplomb" has sometimes been used for unit $A \cdot s^{1/2}$.) Both μ and ϵ can be obtained by computer data-processing of the experimentally generated values of current.

The semiderivative is intermediate between the original function and the normal derivative; the latter can be obtained by applying the semiderivative twice:

$$\frac{d^{1/2}}{dt^{1/2}} \epsilon(t) = \frac{d}{dt} I(t) \tag{16-37}$$

and similarly

$$\frac{d^{-1/2}}{dt^{-1/2}} \mu(t) = \frac{d^{-1}}{dt^{-1}} I(t) \equiv \int_0^t I(t)\,dt \tag{16-38}$$

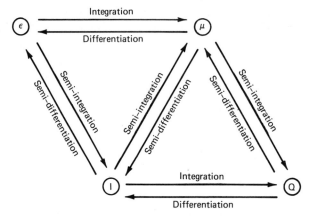

Figure 16-2.   Interrelations of different orders of integration and differentiation, as applied to electrochemical systems.

These relations can be made clearer by the diagrammic scheme in Figure 16-2, relating current and charge [4]. The same result can be obtained from semi-integrating the current as from semidifferentiating the accumulated charge, Q. The independent variable in all cases is time.

It has been shown [4] that for a voltammetric experiment, where potential is changed from well below $E^{\circ\prime}$ for a reducible substance to well above it, that is, from a point where $C_{Ox}^{s} = C_{Ox}^{*}$ to a point at which $C_{Ox}^{s} = 0$, the semi-integral is given by:

$$\mu = nFACD^{1/2} \tag{16-39}$$

This is true regardless of the path by which the potential makes the specified change, such as linear ramp, step, or exponential, and regardless of the degree of reversibility. A plot of μ against E gives a curve with the appearance of a DC polarogram; this has been dubbed a *neopolarogram*.

A related method, introduced in 1975 [5], involves taking the semidifferential, not of the charge, but of the current, thus producing ϵ, which is functionally the derivative of μ:

$$\frac{d\mu}{dt} = \frac{d}{dt}\left(\frac{d^{-1/2}}{dt^{-1/2}}\right)I = \frac{d^{1/2}}{dt^{1/2}}I = \epsilon \tag{16-40}$$

This gives a peak for each reducible species, resembling a differential polarogram. It is shown [6] that for LSV, the peak value of ϵ, designated ϵ_{pk}, is given by:

$$\epsilon_{pk} = \frac{n^2 F^2 AvCD^{1/2}}{RT} \cdot (1.19\,\alpha) \tag{16-41}$$

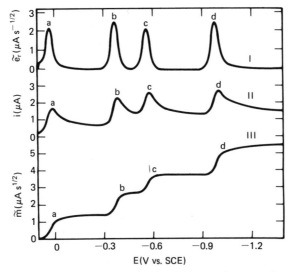

Figure 16-3. Comparison of resolutions of a derivative neopolarogram (I), a voltammogram (II) and a neopolarogram (III) of the same four-component solution containing 50 micromoles each of (a) Cu^{++}, (b) Pb^{++}, (c) Cd^{++}, and (d) Zn^{++}, per liter, in 0.1 M KNO_3. (*Analytical Chemistry* [6].)

where v is the scan rate (in V $\cdot$ s^{-1}). The factor 1.19 appears only for an irreversible reaction. This relation is no longer independent of v nor of reaction kinetics. The peak potential for a reversible process is equal to $E_{1/2}$, but for irreversible systems varies from it in a predictable way. Figure 16-3 shows the relation between the curves of μ, I, and ϵ for a solution containing four reducible species [6].

Measurement of μ and ϵ

An analog circuit element has been designed [7] that will give the semi-integral of a current when connected as the feedback to an op amp, and will give the semi-differential [6] when used in the input to the amplifier. This new component is an *RC*-network that can take several forms, the most used of which is shown in Figure 16-4.

On the other hand, a digital computer can be programmed to accept current values at evenly spaced voltage increments and to compute either the semi-integral or the semidifferential. Considered in this way [8], these techniques appear to be merely variations in signal processing, rather than independent methods, and thus capable of showing little improvement in sensitivity over such methods as differential pulse polarography [8].

Nevertheless, it is quite possible that methods based on the fractional calculus will provide new dimensions to electroanalytical chemistry, due to the lack of dependence on the form of the excitation function.

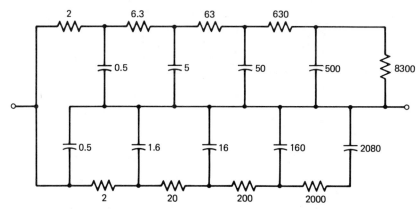

Figure 16-4.   Semi-integrating circuit based on the combination of geometric ladder with its complement.   The component values are in kilohms and nanofarads.   (*Analytical Chemistry* [7].)

A general equation, comparable to Eq. (16-32) can be derived by means of fractional calculus [2, 3].

REFERENCES

1.   K. B. Oldman, *Anal. Chem.*, 1969, *41*, 1904.
2.   K. B. Oldham and J. Spanier, *J. Electroanal. Chem.*, 1970, *26*, 331.
3.   K. B. Oldham and J. Spanier, "The Fractional Calculus," Academic Press, New York, 1974.
4.   M. Grenness and K. B. Oldham, *Anal. Chem.*, 1972, *44*, 1121.
5.   M. Goto and D. Ishii, *J. Electroanal. Chem.*, 1975, *61*, 361.
6.   P. Dalrymple-Alford, M. Goto and K. B. Oldham, *Anal. Chem.*, 1977, *49*, 1390.
7.   K. B. Oldham, *Anal. Chem.*, 1973, *52*, 39.
8.   A. M. Bond, *Anal. Chem.*, 1980, *52*, 1318.

Chapter 17

ELECTRONIC INSTRUMENTATION

The practice of modern electroanalytical techniques is so thoroughly infiltrated with electronics that an understanding of the latter is essential to an intelligent study of the former. Throughout this book frequent reference has been made to electronic instrumentation in the various areas covered but without detailed theory of operation.

This chapter is intended to fill in the background in electronic design and application for those readers whose training in this discipline may need reinforcement. Portions of the material presented are adapted from the authors' *"Analog and Digital Electronics"* [1], and unless otherwise specified, this book can be taken as a general reference.

ELECTRICAL QUANTITIES

The quantities of interest to us are listed in Table 17-1. Some fundamental relations between them are the following:

Ohm's Law:	$E_{dc} = IR$; $E_{ac} = IZ$
Reciprocals:	$G = 1/R$; $Y = 1/Z$
Power Equation:	$P = EI = I^2 R = E^2/R$
Charge:	$Q = \int I \, dt$
Capacitance:	$C = Q/E$
Differential R,Z:	$R = dE_{dc}/dI$; $Z = dI/dE_{dc}$
Differential C:	$C = dQ/dE$

TABLE 17-1
Electrical Quantities

Quantity	Symbol	Unit	Symbol
Potential	E, V	Volt	V
Current	I	Ampere	A
Power	P	Watt	W
Resistance	R	Ohm	Ω
Conductance	G	Siemens[a]	S
Impedance	Z	Ohm	Ω
Admittance	Y	Siemens[a].	S
Quantity	Q	Coulomb	C
Capacitance	C	Farad	F

[a]Formerly, unit mho, symbol, Ω^{-1}.

COMPONENTS

Electronic circuit components are classed as passive or active. *Passive components* cannot increase the power level of a signal. They include resistors, capacitors, inductors, and diodes. Transformers are dual inductors that can change the voltage and current levels of AC signals, but the current that can be supplied always changes inversely to the voltage, maintaining the same nominal power. Both capacitors and transformers can be utilized to provide a low-impedance path for AC signals while maintaining DC isolation.

Diodes are two-terminal semiconductor devices that have the property of passing current in one direction only. They are widely used as rectifiers. Zener diodes differ from others in that a voltage impressed in the normally nonconducting direction will cause the diode to break into conduction at a definite voltage; this is useful in establishing an unchanging potential for reference purposes.

Active components are those that have the ability to increase the power in a signal (though they are not always used in this manner). They must be supplied with an external source of power. The fundamental active component is the *transistor*, which takes many forms. The transistor is a three-terminal semiconductor device, so constructed that a small, varying signal applied to one of its terminals will control the flow of a considerable current between the other two. The common, or bipolar, transistor has a relatively low input impedance, which means that it will draw appreciable current from the signal source. Hence it cannot be used where the source itself is of high impedance; it is, however, inexpensive and robust and serves well in many applications. Field-effect transistors (FET's) have much higher input impedances.

The designers of contemporary instruments are making less and less use of transistors as discrete components. Rather, they employ *integrated circuits* (*IC's*), each of which contains the equivalent of tens to many thousands of transistors fabricated on a single tiny piece of silicon called a *chip*.

OPERATIONAL AMPLIFIERS

The operational amplifier (frequently called "op amp") is one of the most versatile ICs available to the electronic designer. It is a multi-transistor device, provided with two inputs and one output, requiring usually a dual power supply of plus and minus 15 V, DC. Figure 17-1 shows the conventional triangular symbol for an op amp, and a basic circuit using it as the central component. This circuit permits amplification of a signal by a preselected factor, in this case exactly 10.

The basic operation of the op amp is to provide just enough output so that the current through R_f (the feedback path) will force the point marked "SJ" to assume a potential equal to ground (i.e., zero volts). This is caused by the fundamental action of the op amp, which is to maintain actively its two inputs (marked + and -) at the same potential to within a few parts per million. In the circuit shown, the "+" input is grounded, so the potential at SJ will also become equal to ground, even though not directly connected, a condition known as *virtual ground*.

This permits us to draw an important conclusion about the input circuit. By Ohm's law, a voltage impressed across a resistor produces a current through it equal to the ratio E/R. In our circuit, one end of the resistor R_{in} is at a potential of E_{in} volts, while the other end is at (virtual) ground. Hence, a current *must* flow through it. Where does the current actually go? The high input impedance of the amplifier (tens or hundreds of megohms) prevents it from entering within the triangle, so the only place it *can* go is through the feedback resistor, R_f, to the amplifier's output (the apex of the triangle). The nature of the amplifier is such that current can be drawn from it freely without appreciably affecting the output voltage.

This will be clarified if we consider the particular values given in the figure. Let us suppose that $E_{in} = 200$ mV. The current through R_{in}, by Ohm's law, is $0.200/10^4 = 2.00 \times 10^{-5}$ A $= 20.0\ \mu$A. This current, passing through R_f, will

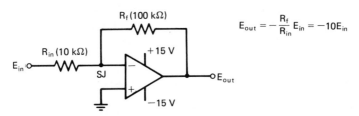

$$E_{out} = -\frac{R_f}{R_{in}} E_{in} = -10 E_{in}$$

Figure 17-1.   An operational amplifier (op-amp) circuit known as an *inverter*. The op amp itself is represented by the triangular symbol. The resistors shown must be supplied as external components. The connections to +15 and –15 volts power supply are usually omitted from diagrams.

produce a potential drop of $R_f I = (2.00 \times 10^{-5})(10^5) = 2.00$ V, which is exactly 10 times the input. It is conventional to consider currents as flowing from a point of positive potential *toward* ground, but in our example the feedback current is flowing *away* from ground and toward the output, so a negative sign must be inserted. The overall operation† can then be described by:

$$E_{out} = -\frac{R_f}{R_{in}} \cdot E_{in} \qquad (17\text{-}1)$$

The circuit discussed above performs the mathematical operation of multiplying a signal voltage by a constant, namely by $(-R_f/R_{in})$. Many other combinations of external components can be deployed around the op amp in place of R_{in} and R_f. In each case, a simple mathematical formula is precisely followed.‡ Figure 17-2 shows several of these op amp circuits, together with their corresponding algebraic descriptives. The circuit in (*a*) is an extension of the previous example, showing that several input voltages can be summed, each amplified by a separate factor. Because the summing junction is at virtual ground, there can be no interaction between the inputs.

The operation of integrating with respect to time can be implemented by the circuit in (*b*) at a rate given by $-E_{in}/RC$.

It is possible to use currents rather than voltages as inputs. This is true of many configurations, an example being given in (*c*), to be compared with Figure 17-1. It gives a voltage output proportional to the current input, and hence is called a current-to-voltage converter. The significance of the negative sign is that a current of electrons coming into the SJ causes a positive output.

Circuits (*d*) and (*e*) of Figure 17-2 differ from the others in that the signal is applied to the "+" (noninverting) input rather than to the summing junction. The advantage of (*d*), in which the output equals the input, is due to the fact that no current is drawn from the source (E_{in}), whereas a considerable current can be taken from the output. This is called a *voltage-follower* or *buffer*; it can also be thought of as a current amplifier. Circuit (*e*) is similar, but permits taking a gain rather than merely reproducing the input; it is known as a *follower-with-gain*.

Two particularly important op amp circuits for electrochemical applications are found in the potentiostat, where a voltage is to be controlled, and in the galvanostat that controls a current. We shall start by considering the galvanostat shown in Figure 17-3. This can be related to the inverter circuit for Figure 17-1, where R_f is replaced by the cell. As before, the current is determined only by the input and is independent of the properties of the cell, including its resistance.

---

†In an alternative way of looking at this circuit, two currents can be discerned, one through R_{in}, the other through R_f, that must be summed to zero at the virtual ground point; this accounts for the designation "SJ" for the Summing Junction, applied to this point.

‡The precision with which a well-designed circuit obeys the formula is limited only by the tolerances of the external components. An accuracy of 0.1 percent is easily obtained. The amplifier itself becomes the limiting factor only at very low voltages or currents, or at high frequency.

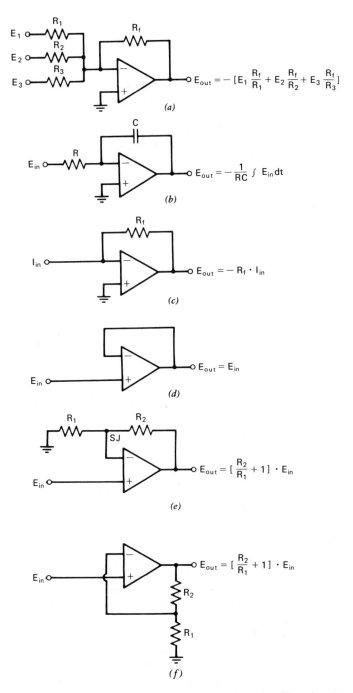

Figure 17-2. An few op-amp circuits with their algebraic formulas. Note that (e) and (f) are identical.

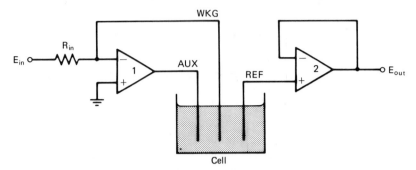

Figure 17-3. A galvanostat circuit. Amplifier #1 is the control element; #2 is a voltage follower to prevent current from passing through the reference electrode.

In this circuit, the input current, determined by the ratio E_{in}/R_{in}, is forced to flow between the auxiliary and the working electrodes, the latter being maintained at virtual ground. The resulting potential, as dictated by the electrochemistry of the cell, is sensed by the reference electrode, which itself draws no current. The signal at E_{in} could be a constant voltage or a ramp, as well as AC, and the current will act accordingly.

A potentiostat is shown in Figure 17-4, in which amplifier #1 controls the potentials of the electrodes in a cell. It acts as a voltage follower, causing the reference electrode to assume the same potential as the input, E_{in}. The working electrode is at virtual ground because of a direct connection to the summing junction of amplifier #2; hence, the difference of potential between working and reference electrodes is equal to E_{in}, as desired. The current that is caused to flow through the working electrode in response to this potential must be supplied by the auxiliary electrode. The high input impedance of amplifier #1 prevents current from passing through the reference electrode. The potential presented to E_{in} from external sources can be any desired voltage function.

In both of these circuits, saturation will occur if the input exceeds the ability of the control amplifier to respond. This ability of the amplifier is called its *com-*

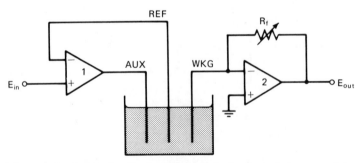

Figure 17-4. A potentiostat circuit. Amplifier #1 is the control element; #2 is a current-to-voltage converter, to ensure that E_{out} is proportional to the current through the working electrode. The value of R_f determines the current sensitivity.

pliance. In the case of the potentiostat, the current compliance is the maximum current that can be supplied without departing from the prescribed voltage. The galvanostat is characterized by its voltage compliance, the greatest voltage that can be exerted to maintain the required current.

Many integrated-circuit op amps are limited to a guaranteed signal output of 10 V, either sign, if powered from a 15 V supply, though some will permit up to about 13 V output. At this maximum output voltage, the current (including the feedback current) that can be taken without deviation from the formula is about 10 to 15 mA. For the majority of op amp applications in laboratory instruments, these restrictions cause no difficulty. If it should be necessary for the output to control larger currents, a *booster* must be used. This may consist of two discrete transistors connected as shown in Figure 17-5, included within the negative feedback loop of the op amp.

One of the great advantages of integrated-circuit op amps is the excellence with which they conform to the theoretical equations. There are limitations, however, that must be kept in mind in addition to the voltage and current restrictions. One is the speed of response; it may take as much as 20 μs to respond to full output, starting at zero. This may seem a short time, but this response time would limit the AC frequency range to about 10 kHz.

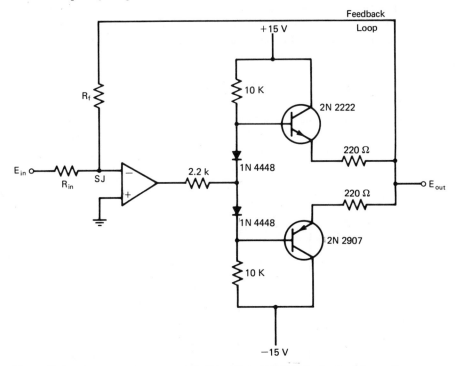

Figure 17-5. A two-transistor current booster, located inside the feedback loop of an op amp. With the values shown, this would provide perhaps 50 mA of current. If signals of one polarity only are expected, one of the transistors can be omitted. The op amp is here configured as a simple inverter, but it could have any of the other usual op amp connections.

Another error that is often important is a voltage offset. Ideally, zero input should give zero output. However, as much as 1 mV, or even more, may appear at the output of common amplifiers. This offset must be eliminated if highest DC precision is required. This can be accomplished either by obtaining a better amplifier or by injecting a corrective voltage upon one of the inputs. Discussion of other possible sources of error can be found in [1].

ANALOG MODULES

The operational amplifiers discussed in the previous section are but one of the many types of analog integrated circuits. The term "analog" refers to a signal that can take any value within its range, in contradistinction to digital signals which are only two-valued.

Among analog components, one somewhat similar to the op amp is the *instrumentation amplifier* shown in Figure 17-6 (amplifier #2). It serves as a high-precision differential amplifier with adjustable gain.

Another module, called a *sample-and-hold amplifier, (S/H)*, permits the temporary retention of some voltage value. It has two modes of operation: SAMPLE, in which the output tracks the input as in a voltage follower, and HOLD, where the output is frozen. In Figure 17-7 is shown an application to sampled polarography. The commands to sample and to hold are digital signals given by a timing circuit.

DIGITAL MODULES

A great variety of digital modules are available, some of great complexity.

Let us consider an application to a sequencer or timer. This unit is usually spe-

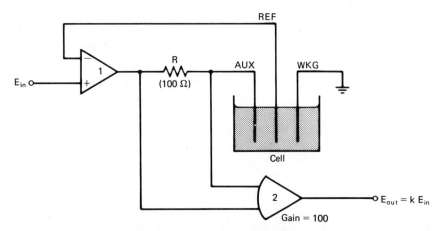

Figure 17-6.   Application of an instrumentation amplifier (#2) as current read-out in a potentiostat. With the values given, a cell current of 1 mA would give a full 10 V output signal.

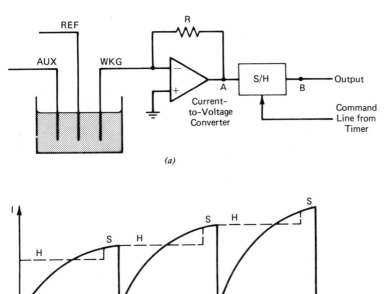

Figure 17-7. Sample-and-hold circuit for a sampling polarograph. (*a*) Schematic. (*b*) Current–time characteristic; the solid line represents the signal at point *A*, the dotted line that at the output (point *B*). The two coincide during the sampling periods, marked *S*.

cifically assembled from a number of standard ICs for the particular application at hand. It provides several outputs that give a pre-established sequence of logic levels shown in Figure 17-8 as a controller for a pulse polarograph. The cycle starts by switching the voltage input to the integrator, thus charging a capacitor for a short time and producing a step in the staircase ramp. The mercury drop is then knocked off, and after a delay of perhaps 2 seconds, the current is sampled by *S/H* #1. Subsequently, the pulse is switched on, and then a second sample is taken during the pulse by *S/H* #2. The instrumentation amplifier gives as its output the difference between the two *S/H* signals. The cycle then repeats.

   The switching operation can be done by means of a solid-state device called a *transmission gate* or *analog gate*. For instance, the control line to the potentiostat might actuate such a gate connecting an additional 50 mV potential to the input, thus producing the desired pulse.

NOISE

Noise can be defined as a spurious signal. Electrical noise is always present to some extent in any circuit, and one of the major thrusts of the instrument designer is to maximize the signal-to-noise ratio, *S/N*. The noise in a circuit is made up of two

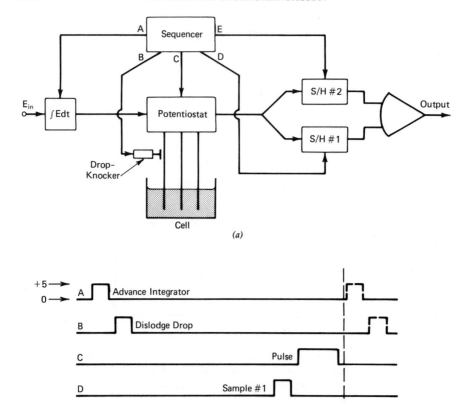

Figure 17-8. The use of a sequencer to control a pulse polarograph. (*a*) Block diagram. (*b*) Timing relations; a second cycle starts at the vertical dashed line.

distinct components. In one of them the power of the noise is independent of the frequency *per se*, but is proportional to the interval of frequencies, Δf, that is being observed. The other type of noise is characterized by a power proportional to the ratio $\Delta f/f$. The former, independent of the frequency, is known as *white noise* (by analogy with white light that contains all frequencies), while the latter is called $1/f$ noise.

White noise originates in part from the random thermal motion of electrons in resistive components; this can be reduced by using components of lower resistance or by cooling them. Noise of the $1/f$ class results from the passage of current through certain types of components including transistors. It can be drastically reduced by *modulating* the desired signal in such a way that it is shifted to higher frequencies.

In addition to noise of internal origin, noise can arise from external sources such as power lines and radio broadcast stations. Noise can also result from mechanical vibrations that cause variation in the capacitance between circuit components. External (environmental) noise can be greatly minimized by careful shielding and grounding of equipment and the use of noise filters on power lines.

For an example of the reduction of external noise by careful grounding, refer to Figure 17-9, which shows the two alternative circuits for current-to-voltage conversion that we have seen before in our discussion of polarographic instrumentation. In (b), the DME is connected to the summing junction of an op amp, whereas in (a) it is grounded. From the noise standpoint, the DME with its elevated mercury reservoir forms an effective antenna for picking up stray signals. In circuit (b), the picked-up noise contributes to the signal going to the recorder, whereas in (a) it is shunted off to ground.

Modulation

As mentioned above, the S/N ratio can be improved by transferring the information from a low frequency or DC to a higher frequency. This constitutes modulation,

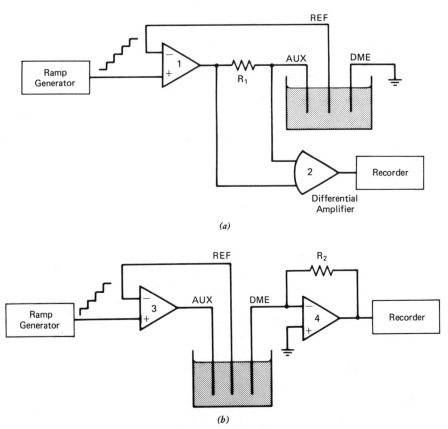

Figure 17-9.   Simplified schematics of two polarographic circuits. (This is identical to Figure 4-12.)

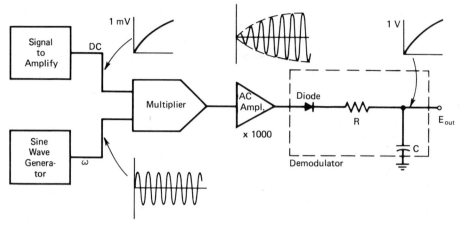

Figure 17-10.   A modulation system.   The analog multiplier acts as a modulator; its output is amplified 1000 times before being demodulated by the diode rectifier.   The final output reproduces the input, but on a scale of volts rather than millivolts.

an example of which is shown in Figure 17-10, where it is effected by means of an *analog multiplier*.   This is an IC that accepts the two inputs E and $A \sin \omega t$, and produces an output given by their product, $E \cdot A \sin \omega t$, an amplitude-modulated sine wave.   The output of the multiplier maintains the frequency ω (the *carrier frequency*) with an amplitude following the signal E.   The modulated carrier is fed into a tuned amplifier, effectively discriminating against noise.   The original signal is recovered by the *demodulator*, consisting of a diode rectifier and low-pass RC filter.

AC polarography shares these noise rejection features, but the situation is more complicated, since the electrochemistry at the electrode is itself affected by the AC modulation.

AC Filters

Since noise extends over all frequencies, while the signal of interest requires only a narrow band, Δf, it is clearly advantageous to eliminate the unwanted frequencies. This can be accomplished by means of *filters*.   The three principal types are shown in Figure 17-11: low-pass, high-pass, and band-pass, corresponding to the desired portion of the frequency spectrum where the information is located.

Another type, the *Twin-T filter*, is shown in Figure 17-12.   This network has a very high impedance for one particular frequency, f_0.   Such a filter can be tuned at the line frequency (60 Hz) to attenuate picked-up noise.   Alternatively, it can be placed in the feedback path of an op amp, resulting in strong amplification at the center frequency and attenuation at other frequencies, both higher and lower.   Its use in both of these connections is illustrated in Figure 17-13, a block diagram of a device intended to amplify a 225-Hz signal while repressing 60 Hz.

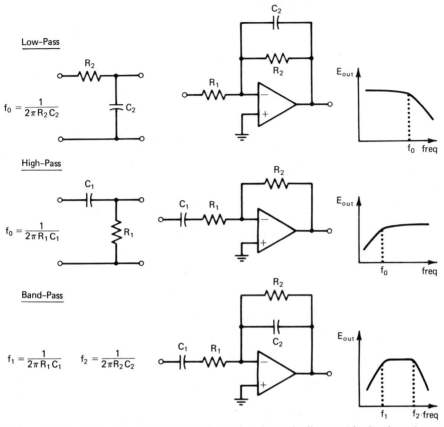

Figure 17-11.   Basic filter circuits in passive and active embodiments (the bandpass is not readily implemented in passive form). The response curves are plotted as the output voltage against the frequency, both on logarithmic scales.

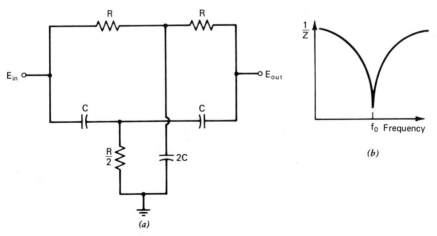

Figure 17-12.   The Twin-T rejection filter. The center frequency is given by $f_0 = 1/(2\pi RC)$. For a sharp resonance, the resistors and capacitors must be closely matched.

241

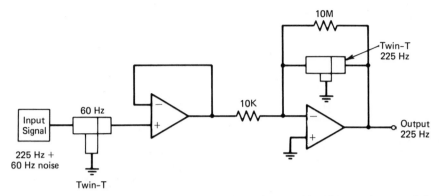

Figure 17-13.   A 225-Hz tuned amplifier with 60-Hz rejection. The 10-MΩ resistor paralleling the 225-Hz Twin-T limits the feedback impedance at resonance so as to avoid saturation of the amplifier.

PHASE RELATIONS

A sine-wave voltage can be described by the expression:

$$E = A \sin (\omega t + \phi) \tag{17-2}$$

where ω is the angular frequency, ϕ is an offset called the *phase*, and A is the amplitude. The phase has no absolute significance, and can only be defined in terms of a reference wave, as shown in Figure 17-14.

The importance of phase relations in electrochemistry is primarily due to the fact that when an alternating voltage is applied to a capacitance, the current that flows *leads* the voltage by $\pi/2$ radians (90°). This applies to the double-layer capacitance at an electrode, just as much as to a discrete capacitor, and must be taken into account in any analysis of phenomena at the electrode.

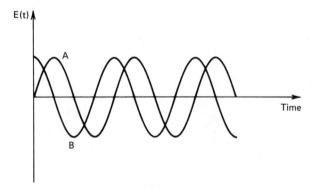

Figure 17-14.   Two sine waves of the same frequency, displaced in time by 1/4 period, or in terms of vectorial angle, by $\pi/4$ radians (90°). Curve *B* reaches a particular point earlier than does *A*, so *B* is said to *lead* and *A* to *lag* in phase.

In discussing AC polarography, it was pointed out that nearly complete rejection of capacitive charging current could be achieved by measuring the current component *in phase with* the applied voltage. This can be implemented by means of a *phase-sensitive detector* (Figure 17-15). The assembly shown including the oscillator, is known as a *lock-in amplifier*. It is widely used wherever it is necessary to recover a low-level AC signal from a noisy background.

In second-harmonic AC polarography, a means must be provided for obtaining a square-wave of precisely twice the frequency of the exciting sine wave. An analog multiplier, together with a clipping circuit, can perform this function easily. The two inputs of the multiplier are both connected to the same sine-wave source, so that the wave is multiplied by itself (squared). The output follows the trigono-

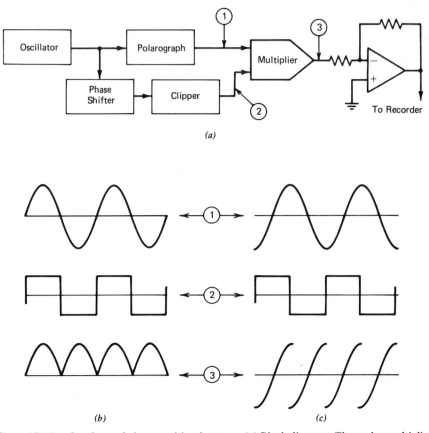

Figure 17-15. One form of phase-sensitive detector. (*a*) Block diagram. The analog multiplier serves to multiply the sinusoidal signal from the polarograph alternatively by +1 and −1. If the square- and sine-wave inputs are in phase (*b*), the negative half-cycles of the sine wave are inverted, whereas the positive portions are not changed. The result is synchronous, full-wave rectification. If the two are 90° out of phase (*c*), the result is a wave form that is half positive, half negative, averaging to zero. The phase-shift module is necessary to achieve synchronism. The clipper converts the sine wave to a square wave.

metric identity:

$$\sin^2 \theta = \tfrac{1}{2} (1 - \cos 2\theta) \tag{17-3}$$

and thus must contain a DC term and a term representing a sine wave of twice the original frequency. The clipper converts the sine wave to a corresponding square wave.

MICROPROCESSORS

A detailed treatment of microprocessors and of computers in general, as applied to electrochemical instrumentation, is beyond the scope of this book. A few remarks about their applicability, however, are in order.

Microprocessors can be used to control an experiment and to process data stemming from it. In a pulse polarograph, for instance, a microprocessor can be programmed to control the timing and magnitude of the pulses, the slope of the DC ramp, the initial and final potentials, the drop time, and the sampling time, all in accordance with commands given through a keyboard. Interlocks can be built into the program to prevent incompatible commands and to detect flaws, such as short-circuited electrodes. The sensitivity of the measuring circuits can also be programmed to match the current being drawn by the cell. In its data processing mode, it can automatically determine and subtract the residual current, identify the peaks, and translate their heights into appropriate concentration units through comparison with standards.

The computer works solely in the digital domain, whereas the electrochemical cell is essentially an analog element. In order to implement control and measurement functions, interfacing between these two must be provided. This requires analog-to-digital (A/D) and digital-to-analog (D/A) converters, both of which are available as discrete plug-in components.

REFERENCES

1. B. H. Vassos and G. W. Ewing, "Analog and Digital Electronics for Scientists," 2nd ed., Wiley-Interscience, New York, 1980.

Appendix 1

SYMBOLS AND ABBREVIATIONS

A	area of electrode surface
a_i	activity of the ith species
C	concentration, general symbol; capacitance
C^*	concentration in the bulk of a solution
C_{dl}	capacitance of the double layer
C^s	concentration at an electrode surface
CV	cyclic voltammetry
D	diffusion coefficient
DME	dropping mercury electrode
DPP	differential pulse polarography
E	potential, general symbol
E°	standard potential relative to the SHE
$E^{\circ\prime}$	formal potential
E^*	potential, referring to bulk of solution
$E_{1/2}$	half-wave potential
ΔE	error potential (in potentiometry); pulse height (in pulse polarography)
E_{AC}	AC potential (root-mean-square unless otherwise stated)
E_{asym}	asymmetry potential
E_{cell}	over-all potential of a cell
E_f	final potential (in LSV or CV)
E_{ind}	potential of indicator electrode
E_i	initial potential (in LSV or CV)
E_{jnct}	liquid junction potential

E_{pk}	peak potential (in LSV or CV)
$E_{pk/2}$	half-peak potential (in LSV or CV)
E_{ref}	potential of reference electrode
E^s	potential, referring to surface conditions
E_s	summit potential (in AC polarography or DPP)
ECM	electrocapillary maximum
F	Faraday constant
f	frequency
f_{mod}	modulation frequency
ΔG	Gibbs free energy
ΔG°	standard Gibbs free energy
ΔG_{chem}	chemical free energy
ΔG_{elchem}	electrochemical free energy
I	current, general symbol
I_0	exchange current
I_{AC}	AC current (root-mean-square unless otherwise stated)
I_{ads}	adsorption current
I_{bkg}	background current
I_{chg}	charging current
I_{cross}	cross-term current
i_d	diffusion current
I_{far}	faradaic current
I_{pk}	peak current (in LSV or CV)
I_s	summit current (in AC polarography or DPP)
IHP	inner Helmholtz plane
IR	ohmic potential drop
J	constant in electrodeposition ($J = AD/\delta$)
k	rate constant, general symbol
k°	heterogeneous rate constant (potentials referred to $E^{\circ'}$)
k_b	rate constant for backward reaction
k_d	diffusion rate constant
k_f	rate constant for forward reaction
k_{hom}	homogeneous rate constant
$k_{pot}^{A,B}$	selectivity coefficient, where B interferes with A
ln	natural logarithm
log	logarithm to the base 10
LSV	linear scan voltammetry
M	molarity
m	molality; rate of flow of mercury

n	number of electrons involved in a redox process
OHP	outer Helmholtz plane
OX	oxidized species; oxidant
pH	negative logarithm of hydrogen-ion activity
PZC	point of zero charge
Q	charge on an electrode
q_i	excess ionic charge
q_m	excess charge on a (metal) electrode
R	resistance, general symbol; universal gas constant
RDE	rotating disk electrode
RED	reduced species; reductant
r	rate of reaction
r_b	rate of backward reaction
r_f	rate of forward reaction
r_t	rate of electron-transfer process
S	slope of pH response curve
SCE	saturated calomel electrode
SHE	standard hydrogen electrode
SMDE	stationary mercury drop electrode
T	Kelvin temperature; transmittance
t	time, general symbol; transference number
Δt	duration of pulse
U	collection efficiency
V	volume, general symbol
V_{det}	detector volume
v	scan rate
Y	admittance, general symbol
Y_{diff}	differential admittance
Z	impedance, general symbol
Z_{diff}	differential impedance
Z_{dl}	impedance of the double layer
Z_{far}	faradaic impedance
z_A	charge on ion of species A
α	transfer coefficient; angle of incidence (in reflection)
Γ	surface excess concentration
γ	activity coefficient; surface tension
δ	Nernst diffusion layer
ϵ	molecular absorptivity; semidifferential function
η_c	concentration overpotential

η_t charge-transfer overpotential

Θ addendum to Nernst equation; time constant (in coulometric analysis)

θ potential variable, $\exp\left[(RT/nF)(E - E_{1/2})\right]$; period (in staircase voltammetry); phase angle between applied voltage and ionic concentration (in AC polarography)

μ chemical potential; semi-integral function

ν kinematic viscosity

ρ density (of mercury)

τ drop time (in polarography); transition time (in thin-layer voltammetry and chronopotentiometry)

Φ flux of matter (in Fick diffusion laws)

ϕ inner potential; phase angle between applied voltage and faradaic current (in AC polarography)

ϕ_2 potential drop across solution part of the double layer

χ surface potential; current function of Nicholson and Shain

ψ outer potential; phase discrepancy angle (in AC polarography)

ω angular frequency, $2\pi f$; rotation rate of RDE

Appendix 2

STANDARD ELECTRODE POTENTIALS

Electrode	Reaction	$E°, V$ versus NHE
F_2, F^-	$F_2 + 2e^- \rightarrow 2F^-$	+2.65
Co^{3+}, Co^{2+}, Pt	$Co^{3+} + e^- \rightarrow Co^{2+}$	+1.82
Au^+, Au	$Au^+ + e^- \rightarrow Au$	+1.68
Ce^{4+}, Ce^{3+}, Pt	$Ce^{4+} + e^- \rightarrow Ce^{3+}$	+1.61
MnO_4^-, Mn^{2+}, Pt	$MnO_4^- + 8H^+ + 5e^- \rightarrow Mn^{2+} + 4H_2O$	+1.51
Au^{3+}, Au	$Au^{3+} + 3e^- \rightarrow Au$	+1.50
Cl_2, Cl^-	$Cl_2 + 2e^- \rightarrow 2Cl^-$	+1.360
$Cr_2O_7^{2-}$, Cr^{3+}, Pt	$Cr_2O_7^{2-} + 14H^+ + 6e^- \rightarrow 2Cr^{3+} + 7H_2O$	+1.33
Tl^{3+}, Tl^+, Pt	$Tl^{3+} + 2e^- \rightarrow Tl^+$	+1.25
O_2, H_2O	$O_2 + 4H^+ + 4e^- \rightarrow 2H_2O$	+1.229
Pt^{2+}, Pt	$Pt^{2+} + 2e^- \rightarrow Pt$	+1.2
Br_2, Br^-	$Br_2(liq) + 2e^- \rightarrow 2Br^-$	+1.065
Hg^{2+}, Hg_2^{2+}, Pt	$2Hg^{2+} + 2e^- \rightarrow Hg_2^{2+}$	+0.920
Ag^+, Ag	$Ag^+ + e^- \rightarrow Ag$	+0.799
Hg_2^{2+}, Hg	$Hg_2^{2+} + 2e^- \rightarrow 2Hg$	+0.789
Fe^{3+}, Fe^{2+}, Pt	$Fe^{3+} + e^- \rightarrow Fe^{2+}$	+0.771
Ag_2SO_4, Ag	$Ag_2SO_4 + 2e^- \rightarrow Ag + SO_4^{2-}$	+0.653
$AgC_2H_3O_2$, Ag	$AgC_2H_3O_2 + e^- \rightarrow Ag + C_2H_3O_2^-$	+0.643
I_2, I^-	$I_2 + 2e^- \rightarrow 2I^-$	+0.536
Cu^+, Cu	$Cu^+ + e^- \rightarrow Cu$	+0.521
Ag_2CrO_4, Ag	$Ag_2CrO_4 + 2e^- \rightarrow 2Ag + CrO_4^{2-}$	+0.446
VO^{2+}, V^{3+}, Pt	$VO^{2+} + 2H^+ + e^- \rightarrow V^{3+} + H_2O$	+0.361
$Fe(CN)_6^{3-}$, $Fe(CN)_6^{4-}$, Pt	$Fe(CN)_6^{3-} + e^- \rightarrow Fe(CN)_6^{4-}$	+0.36
Cu^{2+}, Cu	$Cu^{2+} + 2e^- \rightarrow Cu$	+0.337
UO_2^{2+}, U^{4+}, Pt	$UO_2^{2+} + 4H^+ + 2e^- \rightarrow U^{4+} + H_2O$	+0.334
Hg_2Cl_2, Hg	$Hg_2Cl_2 + 2e^- \rightarrow 2Hg + 2Cl^-$	+0.268
AgCl, Ag	$AgCl + e^- \rightarrow Ag + Cl^-$	+0.222
$HgBr_4^{2-}$, Hg	$HgBr_4^{2-} + 2e^- \rightarrow Hg + 4Br^-$	+0.21
Cu^{2+}, Cu^+, Pt	$Cu^{2+} + e^- \rightarrow Cu^+$	+0.153
Sn^{4+}, Sn^{2+}, Pt	$Sn^{4+} + 2e^- \rightarrow Sn^{2+}$	+0.15
Hg_2Br_2, Hg	$Hg_2Br_2 + 2e^- \rightarrow 2Hg + 2Br^-$	+0.140
CuCl, Cu	$CuCl + e^- \rightarrow Cu + Cl^-$	+0.137
TiO^{2+}, Ti^{3+}, Pt	$TiO^{2+} + 2H^+ + e^- \rightarrow Ti^{3+} + H_2O$	+0.1
AgBr, Ag	$AgBr + e^- \rightarrow Ag + Br^-$	+0.095
UO_2^{2+}, UO_2^+, Pt	$UO_2^{2+} + e^- \rightarrow UO_2^+$	+0.05

Standard electrode potentials (Continued)

Electrode	Reaction	$E°, V$ versus NHE
CuBr, Cu	$CuBr + e^- \rightarrow Cu + Br^-$	+0.033
H^+, H_2	$2H^+ + 2e^- \rightarrow H_2$	0.000
HgI_4^{2-}, Hg	$HgI_4^{2-} + 2e^- \rightarrow Hg + 4I^-$	−0.04
Pb^{2+}, Pb	$Pb^{2+} + 2e^- \rightarrow Pb$	−0.126
Sn^{2+}, Sn	$Sn^{2+} + 2e^- \rightarrow Sn$	−0.136
AgI, Ag	$AgI + e^- \rightarrow Ag + I^-$	−0.151
CuI, Cu	$CuI + e^- \rightarrow Cu + I^-$	−0.185
Mo^{3+}, Mo	$Mo^{3+} + 3e^- \rightarrow Mo$	−0.2
Ni^{2+}, Ni	$Ni^{2+} + 2e^- \rightarrow Ni$	−0.250
V^{3+}, V^{2+}, Pt	$V^{3+} + e^- \rightarrow V^{2+}$	−0.255
$PbCl_2$, Pb	$PbCl_2 + 2e^- \rightarrow Pb + 2Cl^-$	−0.268
Co^{2+}, Co	$Co^{2+} + 2e^- \rightarrow Co$	−0.277
$PbBr_2$, Pb	$PbBr_2 + 2e^- \rightarrow Pb + 2Br^-$	−0.280
Tl^+, Tl	$Tl^+ + e^- \rightarrow Tl$	−0.336
$PbSO_4$, Pb	$PbSO_4 + 2e^- \rightarrow Pb + SO_4^{2-}$	−0.356
PbI_2, Pb	$PbI_2 + 2e^- \rightarrow Pb + 2I^-$	−0.365
Ti^{3+}, Ti^{2+}, Pt	$Ti^{3+} + e^- \rightarrow Ti^{2+}$	−0.37
Cd^{2+}, Cd	$Cd^{2+} + 2e^- \rightarrow Cd$	−0.403
Cr^{3+}, Cr^{2+}, Pt	$Cr^{3+} + e^- \rightarrow Cr^{2+}$	−0.41
Fe^{2+}, Fe	$Fe^{2+} + 2e^- \rightarrow Fe$	−0.440
Ga^{3+}, Ga	$Ga^{3+} + 3e^- \rightarrow Ga$	−0.53
TlCl, Tl	$TlCl + e^- \rightarrow Tl + Cl^-$	−0.557
U^{4+}, U^{3+}, Pt	$U^{4+} + e^- \rightarrow U^{3+}$	−0.61
TlBr, Tl	$TlBr + e^- \rightarrow Tl + Br^-$	−0.658
Cr^{3+}, Cr	$Cr^{3+} + 3e^- \rightarrow Cr$	−0.74
TlI, Tl	$TlI + e^- \rightarrow Tl + I^-$	−0.753
Zn^{2+}, Zn	$Zn^{2+} + 2e^- \rightarrow Zn$	−0.763
TiO^{2+}, Ti	$TiO^{2+} + 2H^+ + 4e^- \rightarrow Ti + H_2O$	−0.89
Mn^{2+}, Mn	$Mn^{2+} + 2e^- \rightarrow Mn$	−1.18
V^{2+}, V	$V^{2+} + 2e^- \rightarrow V$	−1.18
Ti^{2+}, Ti	$Ti^{2+} + 2e^- \rightarrow Ti$	−1.63
Al^{3+}, Al	$Al^{3+} + 3e^- \rightarrow Al$	−1.66
U^{3+}, U	$U^{3+} + 3e^- \rightarrow U$	−1.80
Be^{2+}, Be	$Be^{2+} + 2e^- \rightarrow Be$	−1.85
Np^{3+}, Np	$Np^{3+} + 3e^- \rightarrow Np$	−1.86
Th^{4+}, Th	$Th^{4+} + 4e^- \rightarrow Th$	−1.90
Pu^{3+}, Pu	$Pu^{3+} + 3e^- \rightarrow Pu$	−2.07
AlF_6^{3-}, Al	$AlF_6^{3-} + 3e^- \rightarrow Al + 6F^-$	−2.07
Mg^{2+}, Mg	$Mg^{2+} + 2e^- \rightarrow Mg$	−2.37
Ce^{3+}, Ce	$Ce^{3+} + 3e^- \rightarrow Ce$	−2.48
La^{3+}, La	$La^{3+} + 3e^- \rightarrow La$	−2.52
Na^+, Na	$Na^+ + e^- \rightarrow Na$	−2.714
Ca^{2+}, Ca	$Ca^{2+} + 2e^- \rightarrow Ca$	−2.87
Sr^{2+}, Sr	$Sr^{2+} + 2e^- \rightarrow Sr$	−2.89
Ba^{2+}, Ba	$Ba^{2+} + 2e^- \rightarrow Ba$	−2.90
K^+, K	$K^+ + e^- \rightarrow K$	−2.925
Li^+, Li	$Li^+ + e^- \rightarrow Li$	−3.045

From W. M., Latimer, "Oxidation States of the Elements and Their Potentials in Aqueous Solution," 2d ed., Prentice-Hall, Englewood Cliffs, N.J., 1952.

INDEX

Accuracy, in polarography, 86
AC polarograph, 108
AC polarography, 102
AC pulse polarography, 114
Admittance, 16, 18
Amperometric titrimetry, 211
Amplifiers:
  booster, 235
  instrumentation, 236
  lock-in, 243
  operational, 231
  sample-and-hold, 236
Amplomb (unit), 225
Anode, defined, 2
Anodic oxidations, in polarography, 76
Anodic stripping analysis, 176
Applications of polarography, 86, 87
Asymmetry potential, 45
Auxiliary electrodes, 6

Baseline compensation in polarography, 80
Beer's law, 197
Biamperometric titrimetry, 211
Bipotentiometric titrimetry, 211
Buffers, electronic, 232
Buffers (pH), reference, 46

Calibration curves, 206
Calomel reference electrode (SCE), 63
Capacitance, 18, 25, 185, 230

double-layer, 18, 21, 104, 149, 185
Capacitive current, see Charging current
Capillary, DME, 63, 83
Carbon paste electrodes, 181
Cathode, defined, 2
Cavity potential, 13
Cell constant (conductance), 189
Cells:
  conductance, 184
  polarographic, 80
  thin-layer, 130
Charge-transfer control, 31
Charging current, 25, 95, 109, 118, 145
Charging-current compensation in polarography, 80
Chemical reactions, coupled, 127
Chromatography detector, 173
Chronoamperometry, 35
Chronopotentiometric titrimetry, 214
Chronopotentiometry:
  AC, 146
  constant-current, 140
  varying-current, 147
Class I electrodes, 42
Class II electrodes, 42
Comparison methods, 205
Compensation circuits in polarography, 80
Compliance, 8, 234

Computer control, 244
Conductometric titrimetry,
    212
Conductometry, 184
Constant potential titrimetry,
    214
Convection, methods with, 152,
    162
Conventions, electrochemical,
    8
Cottrell equation, 35, 62,
    255
Coulometers, 159
Coulometric titrimetry, 157,
    215
Coulometry:
    controlled current, 157
    controlled potential, 156
Coulostatic analysis, 149
Coulostatic titrimetry, 214
Current compliance, 8, 235
Current density, 25
Current measurements, 25, 232
CV, 118
Cyclic voltammetry, 118

Depolarized electrode, 20
Detector, phase-sensitive, 108, 110,
    243
Differential pulse polarography (DPP),
    93
Diffusion:
    finite, 130
    in voltammetry, 61
Diffusion coefficient, 27, 219
Diffusion control, 31, 218
Diffusion current, 71
Diffusion layer:
    in electrodeposition, 154
    in hydrodynamic voltammetry, 164
    in LSV, 122
Diffusion transport, 33, 218
Disk electrodes, rotating, 163
Distortion, harmonic, 109
DME, 62
Double layer, 18, 20
DPP, 93
Drop-knocker, 80, 100
Dropping mercury electrode (DME), 62
Drop time, in polarography,
    71
Dummy cell, 2
Dynamic range, 206

ECL, 202
ECM, 24
Eisenman equation, 47
Electrical double layer, 18, 20
Electrocapillarity, 23
Electrocapillary maximum, 24
Electrochemical conventions, 8
Electrochemical measurements, 12
Electrochemical photochemistry,
    199
Electrochemiluminescence, 202
Electrodeless cells, 187, 193
Electrodeposition, 152, 177
Electrodes:
    auxiliary, 6
    carbon paste, 181
    chemically modified, 136
    classes of, 42
    counter, 6
    depolarized, 20
    dropping mercury, 62
    flow-through, 167
    glassy carbon, 181
    hanging mercury drop, 180
    hydrogen, 9, 38
    indicator, 6, 41
    membrane, 44, 48, 50, 136
    mercury drop, 180
    mercury film, 181
    minigrid, 197
    platinized, 184
    polarized, 20
    reference, 6, 39
    ring-disk, 171
    rotating disk, 163
    semiconductor, 201
    tubular, 167
    wall-jet, 174
    WIG, 181
    working, 6
Electrolyte, supporting, 22, 69
Electrolytic cell, 3
Electromotive force (emf), 12
Electronics, 229
Enzyme electrodes, 56
Equivalent circuit, 2
Equivalent conductance, 189
Error function (erf), 131
Exchange current, 29

Faradaic impedance, defined, 18
Faradaic rectification polarography,
    113

Faraday constant, 3, 156
Faraday's law, 155
Fick's laws, 33, 122, 143, 166, 219, 220
Filters, electrical, 240
Finite diffusion voltammetry, 130
Flow-through electrodes, 167
Flux, 218
Formal potential, 9, 41
Fractional calculus, 225
Free energy, 3

Galvanic cell, 3
Galvani potential, 13
Galvanostat, 6, 158, 232
Gates, analog (transmission), 237
Glass electrode (pH), 44
Glassy carbon electrodes, 181

Half-cell potential, 9
Half-wave potential, 74, 83, 85
Harmonic distortion, 109
Helmholtz planes, 21
Heterogeneous rate constant, 28
Heyrovský-Ilkovič equation, 73, 225
Hydrodynamic voltammetry, 162

Ilkovic equation, 70
Immobilized reagent voltammetry, 135
Impedance, 16, 18, 102
Indicator electrodes, 6, 41
Inner Helmholtz plane, 21
Inner potential, 13
Instrumentation amplifiers, 236
Integrators, 159, 232
Intermetallic compounds, 182
Intermodulation polarography, 113
Ionic equivalent conductance, 189
Ionic mobility, 189
Ion-selective electrodes, 48
Ion-sensitive FETs (ISFETs), 56
IR-drop, 15
Irreversible process, 27, 75
Irreversible systems:
  in LSV, 126
  in polarography, 74
ISFETs, 56

Junction potential, liquid, 14, 37, 39

Karaoglanoff equation, 144, 225

Levich equations, 167, 172
Linear sweep voltammetry, 116
Linear systems, defined, 17
Liquid junction potential, 14, 37, 39
Lock-in amplifiers, 243
LSV, 116
Luggin capillary, 7

Magnetic induction, 187
Matrix effects, 206
Maximum:
  electrocapillary, 24
  polarographic, 78
Maximum suppressors, 78
Mediators, chemical, 199
Membrane electrodes, 44, 50
Mercury film electrodes, 181
Microprocessors, 244
Migration, in voltammetry, 61
Minigrid electrodes, 197
Mobile-carrier electrodes, 53
Mobility, ionic, 189
Modulation:
  AC, 102, 146, 169, 239
  frequency, 170
  hydrodynamic, 169
Multiplier, analog, 240

Neopolarograms, 226
Nernst equation, 9, 41
Nernst layer, see Diffusion layer
Nerstian slope, defined, 45
Noise, electrical, 237
Normal pulse polarography, 93

Ohm's law, 7, 229
Ohmic drop, 15
Operational amplifiers, 231
Optical methods, 195
Optically transparent electrodes, 195
Organic applications of polarography, 87
Oscillographic polarography, 116
Oscillometer, 193, 213
OTEs, 195
Outer Helmholtz plane, 21
Outer potential, 13
Overpotential:
  charge-transfer, 15
  concentration, 15
Overvoltage of hydrogen, 33
Oxygen sensor, Clark, 138

pH-meter, 45, 58
pH-stat, 215
Phase (Gibbs), 13
Phase (in alternating current), 105, 242
Phase-sensitive detection, 108, 110, 243
Photochemistry, 199
Photoelectrochemistry, 201
Photoelectrodes, 199
Platinization, 184
Point of zero charge (PZC), 24
Polarized electrode, 20
Polarograph:
  AC, 108
  DC, 78
  pulse, 100
  second-harmonic AC, 112
Polarographic maxima, 78
Polarography, 62
  AC, 102
  AC-pulse, 114
  DC, 62
  differential pulse, 93
  Faradaic rectification, 113
  intermodulation, 113
  normal pulse, 93
  oscillographic, 116
  pulse, 90, 93
  second harmonic AC, 109
  square-wave, 92
  tast, 79
Potentiometer, 5
Potentiometric titrimetry, 209
Potentiometry, 4, 37
Potentiostat, 7, 158, 234
Precision in polarography, 86
Princeton Applied Research Corporation, 67
Pseudo-bridge, 191
Pulse polarography, 90
  AC, 114
  differential, 93
  normal, 93
PZC, 24

Quasi-reversible process, 27, 75

Raman spectroscopy, 203
Randleš-Sevčik equation, 124, 179
Rate constant, heterogeneous, 28
RDE, 163
Redox, 1

Redox electrodes, 42
Reference buffers (pH), 46
Reference electrodes, 6, 39
Residual current, 80
Resistance compensation in polarography, 80
Resolution in polarography, 84
Reversible process, 27, 32, 75
Ring-disk electrode, 171
Rohm and Haas Company, 78
Rotating disk electrode, 163, 171

Sample-and-hold amplifiers, 236
Sampling circuit, 63, 79
Sand equation, 144, 225
Sargent-Welch Scientific Company, 78
SCE, 63
Second-harmonic AC polarography, 109
Selectivity coefficient, 47
Semi-differential techniques, 225
Semi-integral techniques, 225
Semiconductor electrodes, 201
Sensitivity in DC polarography, 86
Sensitized electrodes, 54
SHE, 9, 38
Signal-to-noise ratio, 102, 146, 206
SMDE, 67
S/N ratio, 102, 146, 206
Sparging, 63
Spectroelectrochemistry, 195
Square-wave polarography, 92
Staircase ramp:
  in LSV, 128
  in polarography, 63, 78
  in pulse polarography, 99
Standard addition, 207
Standard hydrogen electrode, 9, 38
Standard potential, 9
Standard solutions, 205
Standard subtraction, 208
Stationary mercury drop electrode, 67
Stripping analysis, 176
Summit, in AC polarography, 103
Supporting electrolyte, 22, 69, 83
Surface excess, 120

Tafel plot, 30
Tast circuit, 79

Thin-layer cells, 130
Titrimetry, 208
  amperometric, 211
  biamperometric, 211
  bipotentiometric, 211
  chronopotentiometric, 214
  conductometric, 212
  constant potential, 214
  coulometric, 157, 215
  coulostatic, 214
  oscillometric, 213
  potentiometric, 209
Tortuosity factor, 136
Transfer coefficient, 28
Transference numbers, 40
Transient, defined, 26
Transition time:
  in chronopotentiometry, 141
  in thin-layer cells, 134
Transport control, 31
Triton X-100, 78

Tubular electrodes, 167

Underpotential, 178

Voltage compliance, 8, 235
Voltage follower, 5, 232
Voltage measurements, 12
Voltammetry:
  cyclic, 118
  defined, 60
  finite diffusion, 130
  hydrodynamic, 162
  immobilized reagents, 135
  linear sweep, 116
  reverse-scan, 179
Volta potential, 13

Wall-jet electrode, 174
Wheatstone bridge, AC, 190
WIG electrodes, 181
Working electrodes, 6